THE **BOTANY**
COLORING BOOK

The
BOTANY
COLORING
BOOK

by Paul G. Young

Illustrations by
Jacquelyn Giuffré

BARNES & NOBLE BOOKS
A DIVISION OF HARPER & ROW, PUBLISHERS
New York, Cambridge,
Philadelphia, San Francisco, London,
Mexico City, São Paulo, Sydney

Paul G. Young, a biologist and botanist, is a lecturer at the City College of San Francisco and botanist for the William Joseph McInnes Memorial Botanical Garden at Mills College, Oakland, California.

Jacquelyn Giuffré has worked as a natural science illustrator for the past four years. She has illustrated a book on trees for the Huntington Library in San Marino, California, and was the illustrator for the **Pacific Coast Bird Finder.**

This book was produced by Coloring
    Concepts Inc.
P.O. Box 324, Oakville, Ca. 94562

The book editor was Joan Elson
Layout was by Wynn Kapit
The copy editor was Sylvia Stein
Type was set by Ampersand Design,
    San Francisco
Page makeup and production coordination
    was by Donna Davis
The proofreader was Sue Gamlen

FIRST EDITION

ISBN: 0-06-460302-4

82   83   84   85   10
9   8   7   6   5   4   3   2   1

To my family for their unwavering love and support, and especially to my son, Brian, for his continual patience.

Also, to all people who, through their enjoyment and appreciation of nature, have come to realize there is far more to success and happiness than power and wealth.

—Paul Young

To my sister Nancy Broaders, for her continual support, and to my husband Craig for his love and patience.

—Jacque Giuffré

TABLE OF
CONTENTS

PRODUCER'S
ACKNOWLEDGMENT

Coloring Concepts had tremendous support during the development of this and the other three books of this initial Series (Botany, Human Evolution, Marine Biology, and Zoology). The people at Harper and Row, especially Irv Levey and Tom Dorsaneo, have encouraged us, responded to our needs, and demonstrated patience and understanding in our struggle to make a deadline. The following friends have directly aided our endeavor: Howard Nemerovski, Tom Larsen, Stu Boynton, Kathy Dahl, Bill and Danielle Brown, Don Jones, Terry Anderlini, John Moran, and Dr. Jack Lange. Their support and confidence in us is greatly appreciated. Dr. Lange (Lange Medical Publications) has given us advice and direction as we develop as publishers. His philosophy of publishing has been a source of inspiration to us. We are much indebted to Donna Davis who coordinated our book productions and the fine people who worked with her: Libbie Schock and Nancy Steele at Ampersand Design, Sue Gamlen, Bob Nemerovski, and our copy editors, Sylvia Stein and David Cross. The concern and reassurance of our friends Gene Mattingly, Regan Anderlini, Ken and Ute Christensen, Ellis and Merilyn Bowman, Maurice and Ann McColl, Bob and Nancy Teasdale, Sharon Boldt, and Julie and Nancy at Schoonmaker provided the background so necessary for us to achieve our goal. We are very grateful.

Joan Warrington Elson
Lawrence M. Elson

July 1981

PREFACE

I hear, and I forget;
I see, and I remember;
I do, and I understand.
　　　　　—*Old Chinese Proverb*

The Botany Coloring Book is designed to be an effective learning aid for the study of plant structures, functions, and life histories. No previous botanical or biological knowledge is required for use of this book. Because of its unique and effective approach to the study of plants, colorers with various backgrounds, from high school level through college level, as well as persons with strictly avocational interests, will find *The Botany Coloring Book* to be a useful tool for furthering their knowledge and understanding of plants. Learning plant structures, functions, and life histories is clarified and simplified through the coordination of reading and coloring. This combination of activities by the colorer provides significant learning reinforcement and retention enhancement. To obtain the most effective results from *The Botany Coloring Book,* the introductory guide to coloring and color notes should be carefully followed.

The Botany Coloring Book consists of 100 units. The first sixteen units contain introductory material about plant cell structure, cell division, simple growth, asexual and sexual reproduction, and the general plant life history cycles. These units should be done in sequence and well understood before work on any of the plant life history units is at-tempted. Within a plant group (e.g. flowering plants), the units should be done in sequence.

The scope of this book is broad, but the pages are few. Therefore, an attempt was made to cover important aspects of introductory botanical study. Emphasis is placed on structures, functions, and life histories of representative members of major plant groups, and on general plant function. While using this book, the colorer should continually attempt to make connections between the general habitat of a plant group and how it is adapted for successful survival under the conditions of its habitat type.

In addition to the support provided by my family and friends, I am indebted to Dr. Cherie Wetzel and Dr. John Thomas for reviewing the manuscript and providing many useful suggestions. Audrey Teasdale also reviewed a portion of the book.

Dr. Wetzel is responsible for my introduction to Dr. Larry Elson who suggested doing *The Botany Coloring Book.* Professional, technical and artistic assistance was provided by a number of qualified people, but assistance above and beyond the call of duty was given by Joan Elson in editorial assistance and by Wynn Kapit in design and artistic assistance. Many thanks go to the artist, Jacque Guiffré, who did a fine job and has a special talent for botanical illustration. Elly Simmons illustrated Plates 4, 6 and 98 for which I am grateful.

　　　　　　　　　　　　　　　Paul G. Young

July 1981
Oakland, Ca.

HOW TO USE THIS BOOK
COLORING INSTRUCTIONS

1. This is a book of illustrations (plates) and related text pages in which you (the colorer) color each structure indicated the same color as its name (title), both of which are linked by identical numbers (subscripts). In the doing of this, you will be able to relate identically colored name and structure at a glance. Structural relationships become apparent as visual orientation is developed. These insights, plus the opportunity to display a number of colors in a visually pleasing pattern, provide a rewarding learning experience.

2. You will need coloring instruments. Colored pencils or colored felt-tip pens are recommended. Laundry markers (with waterproof colors) and crayons are not recommended: the former because they stain through the paper, and the latter because they are coarse, messy and produce unnatural colors.

3. The organization of illustrations and text is based on the author's overall perspective of the subject and may follow, in some instances, the order of presentation of a formal course of instruction on the subject. To achieve maximum benefit of instruction, you should color the plates in the order presented, at least within each group or section. Some plates may seem intimidating at first glance, even after reviewing the coloring notes and instructions. However, once you begin coloring the plate in order of presentation of titles and reading the text, the illustrations will begin to have meaning and relationships of different parts will become clear.

4. As you come to each plate, look over the entire illustration(s) and note the arrangement and order of titles. Count the number of subscripts to find the number of colors you will need. Then scan the coloring instructions (printed in bold face type) for further guidance. Be sure to color in the order given by the instructions. Most of the time this means starting at the top of the plate with (A) and coloring in alphabetical order. Contemplate a number of color arrangements before starting. In some cases,

you may want to color related forms with different shades of the same color; in other cases, contrast is desirable. In cases where a natural appearance is desirable, the coloring instructions may guide you or you may choose colors based on you own knowledge and observations. One of the most important considerations is to link the structure and its title (printed in large outline or blank letters) with the same color. If the structure to be colored has parts taking several colors, you might color its title as a mosaic of the same colors. It is recommended that you color the title first and then its related structure. If the identifying subscript lies within the structure to be colored and is obscured by the color used, you may have trouble finding its related title unless you colored it first.

5. In some cases, a plate of illustrations will require more colors than you have in your possession. Forced to use a color twice or thrice on the same plate, you must take care to prevent confusion in identification and review by employing them on separate areas well away from one another. On occasion, you may be asked to use colors on a plate that were used for the same structure on a previous related plate. In this case, color their titles first regardless of where they appear on the plate. Then go back to the top of the title list and begin coloring in the usual sequence. In this way, you will be prevented from using a color already specified for another structure.

6. Symbols used throughout the book are explained below. Once you understand and master the mechanics, you will find room for considerable creativity in coloring each plate. Now turn to any plate and note:

   a. Areas to be colored are separated from adjacent areas by heavy outlines. Lighter lines represent background, suggest texture, or define form and (in the absence of "don't color" symbols) should be colored over. If the colors you used are light enough, these texture lines may show through, in which case you may wish to draw darker or heavier over these lines to add a three-dimensional effect. Some boundaries between coloring zones may be represented by a dot or two or dotted lines. These represent a division of names or titles and indicate that an actual structural boundary may not exist or, at best, is not clearly visible.

b. As a general rule, large areas should be colored with light colors and dark colors should be used for small areas. Take care with very dark colors: they obscure detail, identifying subscripts, and texture lines or stippling. In some cases, a structure will be identified by two subscripts (e.g., A + D). This indicates you are looking at one structure overlying another. In this case, two light colors are recommended for coloring the two overlapping structures.

c. Any outline-lettered word followed by a small capitalized letter (subscript) should be colored. In most cases, there will be a related structure or area to color. If not, the symbol N.S. (not shown) will follow the word; or, the word functions as a heading or sub-heading and is colored black ($\bullet$) or gray ($\star$). Outline titles (headings) with no subscript following are to be left uncolored.

d. In the event structures are duplicated on a plate, as in left and right parts, branches, or serial (segmented) parts, only one may be labeled with a subscript. Without boundary restrictions or instructions to the contrary, these like structures should all be given the same color.

e. In looking over a number of plates, you will see some of the following symbols:

$\bullet$ = color black; generally reserved for headings/subheadings

$\star$ = color gray; generally reserved for headings/subheadings

$-|-$ = do not color

A( ) = set next to titles subscript; signals this structure composed of parts listed below with same letter but different exponents; receives same color; only its parts are labeled in illustration

A^1, A^2, etc.= identical letter with different exponents implies parts so labeled are sufficiently related to receive same color

N.S. = not shown

= mitosis*

= meiosis*

= syngamy or fertilization*

*It is suggested that you select a separate color for each of the symbols shown above and use them throughout the book wherever these symbols appear.

7. In the text, certain words are set in *italics*. According to convention, the generic name and species of an animal or plant are set this way (e.g. *Homo sapiens*). In addition, the title of any structure to be colored on the related (facing) plate is set in italics (except for headings and sub-headings). This is to enable you to quickly spot in the text the title of a structure to be colored.

THE **BOTANY**
COLORING BOOK

1
IMPORTANCE OF PLANTS TO HUMANS

The importance of plants to the survival of humans and other animals cannot be overemphasized. Our basic needs of food, shelter, clothing, and oxygen are provided, either directly or indirectly, by plants.

Color the leaf and diagram labeled "construction" (A).

Plants provide strong yet lightweight materials used in *construction*. Many cultures rely upon plants to provide all or part of the materials used in shelter *construction*. In some types of shelters, such as yurts and teepees, poles obtained from plants are used as a basic framework, which are covered by animal skins or plant materials such as palm leaves. In other types of shelters, such as wood frame houses, wood is the major *construction* material.

Color the leaf and diagram labeled "energy" (B).

Light from the sun is the major source of *energy* for the earth. Through photosynthesis, plants capture light *energy* and store it as organic chemical *energy*. We utilize this stored *energy* internally through the consumption of food as a cellular *energy* source and externally through the combustion of organic materials, such as wood, as a heat or light *energy* source.

Color the leaf and diagram labeled "food" (C).

All animals, including humans, are heterotrophic. That is, they cannot manufacture their own *food* from inorganic substances. Instead, they must rely upon organic materials as their *food*, or internal energy, source. Most plants are autotrophic. That is, they utilize an inorganic energy source, such as the sun, to manufacture energy-rich organic compounds from inorganic substances. We grow and harvest plants for direct consumption as *food*, such as cereals, fruits, and vegetables, or indirectly as in the fodder and grains fed to livestock which are consumed as *food*.

Color the leaf and diagram labeled "oxygen" (D).

During the energy-capturing, food-producing process of photosynthesis, green plants split water and release free *oxygen* into the atmosphere. This is the source of the atmospheric *oxygen* we breathe. Without the continual release of large quantities of *oxygen* by green plants through photosynthesis, the free *oxygen* in the atmosphere would be depleted by the energy-releasing processes, called respiration, of living organisms and by combining with inorganic and dead organic materials through chemical oxidation (as in the rusting of iron and burning of wood).

Color the leaf and diagram labeled "food processing" (E).

Some plants are very important in certain types of *food processing*. Yeast, a microscopic fungus, is important to us for baking and producing alcoholic beverages. For example, wine yeasts utilize the natural sugar in grapes as an energy source (yeasts are heterotrophic). In doing so, carbon dioxide gas, CO_2, the fizz in champagne and beer, is released, and alcohol is produced as a by-product. The increasing alcohol content eventually kills the yeast and preserves the beverage from spoilage. In yeast bread baking, the CO_2 released by the yeast during fermentation of sugar in the dough creates a dough filled with minute gas bubbles that expand during baking to produce a light, airy product.

Some bacteria are also important in *food processing*. The manufacture of many dairy products, including cheeses, yogurt, and sour cream, is based upon the activities of bacteria that utilize milk as their food source. Enzymes secreted by the bacteria alter the milk, and by-products toxic to the bacteria eventually kill them and produce a storable, preserved food.

Color the leaf and diagram labeled "clothing" (F).

Some plants produce abundant amounts of long, thin cells called fibers that, because of their length and strength, are useful for weaving. In much of the world, the cotton plant, *Gossypium*, is the major source of plant fibers used to manufacture *clothing*. Other plants, such as flax *(Linum)* and hemp *(Cannabis)*, provide coarse fibers.

Color the leaf and diagram labeled "medicine" (G).

Penicillin, an antibiotic *medicine* obtained from a fungus (mold), was "discovered" in the mid 1940s. Quinine, used to alleviate malaria symptoms, is obtained from the bark of the *Cincona* tree. Cascara *(Rhamnus)* bark is the source of an effective laxative. Birth control drugs were first obtained from a wild yam, *Dioscorea*, and morphine is derived from opium produced by the opium poppy, *Papaver somniferum*.

IMPORTANCE OF PLANTS TO HUMANS.

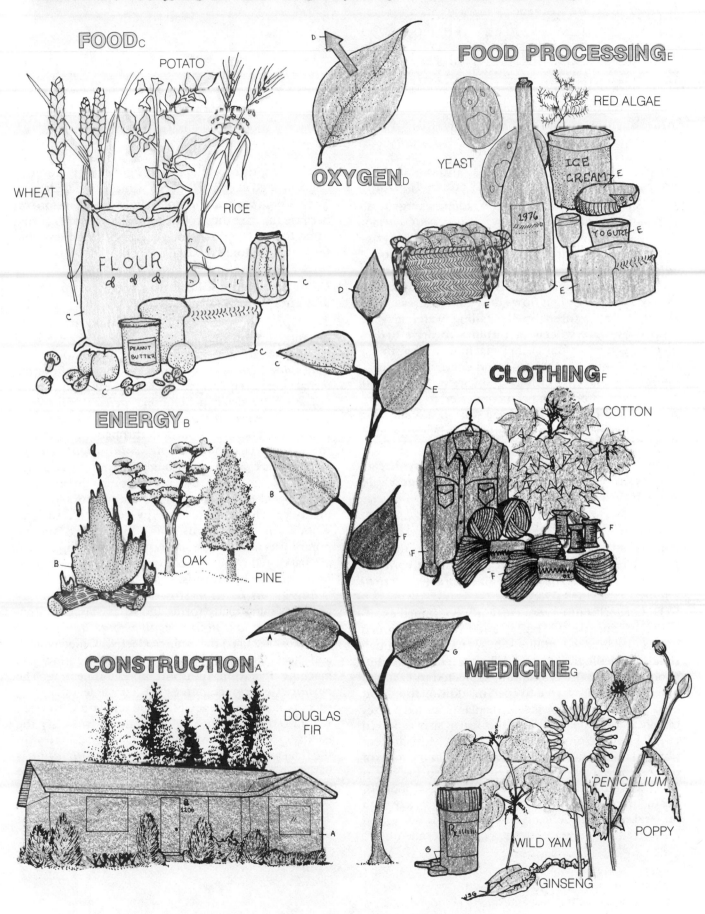

FOODc

POTATO

WHEAT

RICE

FLOUR

PEANUT BUTTER

OXYGEND

FOOD PROCESSINGe

RED ALGAE

YEAST

1976

ICE CREAM

YOGURT

CLOTHINGF

COTTON

ENERGYB

OAK

PINE

CONSTRUCTIONA

DOUGLAS FIR

MEDICINEg

PENICILLIUM

WILD YAM

POPPY

GINSENG

2
TAPPING EARTH'S MAJOR ENERGY RESOURCE

Color solar heat energy (A) and solar heat energy resources (B). Use light colors for both.

Earth's primary energy source is radiant energy from the sun, called solar energy. *Solar heat energy,* through its influence in such energy transfer as water evaporation and air movement, creates a variety of *solar heat energy resources.* For example, evaporated water forms atmospheric moisture, eventually falls as rain or snow, and then flows downhill as streams and rivers. Energy stored by the falling water may be tapped by waterwheels or turbines to drive hydroelectric generators and other machinery. In a similar manner, windmills may tap wind energy. Both these energy transducers (turbines and windmills) tap *solar heat energy* after it has been converted to a different form. The increasingly popular solar heat collecting panels directly tap *solar heat energy* by heating liquid.

Color solar light energy (A1), photosynthesis (C), chemical energy (D), and nonrenewable solar light energy resources (E).

Most solar energy arrives in the form of *solar light energy,* which, with the exception of solar electric panels, can be captured only by green plants through *photosynthesis,* the biological process that converts unstorable *solar light energy* into storable *chemical energy.*

The majority of energy currently consumed externally, or outside the body, is derived from green plants that existed millions of years ago. Through time and geologic changes, the ancient plants (left panel of the leaf) were buried, subjected to heat and pressure, and converted to coal. In addition to ancient land plants, microscopic marine plants, called phytoplankton (middle panel of the leaf), also captured large amounts of *solar light energy.* Most ancient phytoplankton stored *chemical energy* in the form of oil. As these plants died, they settled to the ocean floor and collected in depressions. These pools of dead, oil-rich plants were eventually covered with mineral sediment, subjected to heat and pressure, and converted to crude oil. Thus, coal and crude oil are forms of captured ancient *solar light energy,* stored as concentrated *chemical energy* waiting to be released. They are *nonrenewable solar light energy resources* because they are the product of *solar light energy* that came to Earth millions of years ago, to an environment that no longer exists.

Color renewable solar light energy resources (F) and energy utilization (G).

The capture of *solar light energy* has occurred without interruption since green plants first appeared on Earth. Though we presently depend upon *nonrenewable solar light energy resources* for most of our external energy requirements, such as heating and transportation, we also depend upon *renewable solar light energy resources* for both externally and internally consumed energy resources, through ingestion of food materials. *Renewable solar light energy resources* are produced through the continual capture of *solar light energy* by contemporary green plants (right panel of the leaf). Some contemporary green plants provide external energy resources, such as wood, which may be used as a fuel. Other contemporary green plants, such as the various crop plants, are harvested either for consumption by livestock or for direct use by humans.

Energy utilization to satisfy human energy demands takes many forms and is lavishly consumptive. Most of the *nonrenewable solar light energy resources* are used to produce electrical, mechanical, and heat energy to meet needs for lighting and machine operation, transportation, and heating. The *chemical energy* resources we consume internally, such as meat and vegetable products, are converted within our bodies into cellular energy to satisfy the energy needs of growth and metabolic activities. In the final analysis, the bulk of our *energy utilization* is based on energy resources that are derived from the past and present conversion of *solar light energy* into *chemical energy* by *photosynthesis;* and use that exceeds production, which is the current status, cannot continue indefinitely.

SOLAR ENERGY CHAIN.

SOLAR HEAT ENERGY_A
SOLAR HEAT ENERGY RESOURCE_B
SOLAR LIGHT ENERGY (SLE)_{A¹}
PHOTOSYNTHESIS_C
CHEMICAL ENERGY_D
NONRENEWABLE (SLE) RESOURCE_E
RENEWABLE (SLE) RESOURCE_F
ENERGY UTILIZATION_G

3
THE PLANT CELL

With few exceptions, all organisms have the cell as their basic unit of organization. Plants range from microscopic unicellular to huge multicellular structures composed of billions of cells. Because of cellular specialization, many variations in cell structure, composition, and function exist. The cell illustrated provides a generalized overview of plant cell structure.

Color the cell wall (A), middle lamella (B), plasmodesmata (C), and plasma membrane (D).

One unique feature of plant cells is the *cell wall,* a porous, nonliving, supportive, and protective shell. The *cell wall* is constructed by the protoplast, the living part of the cell. In multicellular plants, *cell walls* of adjacent cells are in contact and held together by the *middle lamella,* a thin cementing layer. In addition, minute pores, the *plasmodesmata,* traverse adjacent *cell walls* and provide a living interconnection between protoplasts. The protoplast contains numerous membrane-bound, organized structures collectively called organelles and is delimited by the *plasma membrane,* or plasmalemma, a single membrane closely appressed to the *cell wall.* Unlike the *cell wall,* the *plasma membrane* is not freely porous to substances. Instead, it mediates the movement of substances into and out of the protoplast by selective control.

Color the hyaloplasm (E), using a light color, and the nucleus (F) and nucleolus (G). Because the plastid (H) illustrated is a chloroplast, color it a shade of green.

The major components of the protoplast are the *plasma membrane,* cytoplasm, and *nucleus.* Most cellular metabolic activity occurs within the cytoplasm. The cytoplasm consists of various organelles and the viscous matrix, the *hyaloplasm,* in which they are suspended. The *nucleus,* a large structure surrounded by *hyaloplasm,* contains the genetic material that determines the cell's structure and function. Within the *nucleus* is a small, spherical *nucleolus,* which is a control center for the production of ribonucleic acid (RNA). Some *nuclei* have as many as four *nucleoli.* The major cytoplasmic organelles of plant cells are *plastids, mitochondria, dictyosomes, vacuoles,* and *ribosomes,* though other types are also present.

One to numerous *plastids* may be present within a plant cell. *Plastids* contain pigments or stored food materials or a combination of both. Several kinds of *plastids* are recognized based on their internal structure and content. The *plastid* type illustrated is a chloroplast that is responsible for the green color of plants and the capture of light energy.

Color the mitochondria (I), dictyosome (J), vesicle (J1), vacuole (K), tonoplast (K1), and crystal (L).

The *mitochondrion* is a much smaller organelle than the *plastid. Mitochondria* release energy from food materials for use in cellular functions. *Dictyosomes,* layers of flattened, membrane-bound sacs, function in the formation and encapsulation of various materials. Membrane-bound *vesicles* pinch off from the *dictyosomes* and are transported to areas where their contents are used. In most plants, one large *vacuole* occupies most of the volume of a mature cell. *Vacuoles* are bound by a single membrane, the *tonoplast,* and contain an aqueous solution in which many substances are dissolved. Some substances are commonly held in a concentrated *crystal* form. *Vacuoles* function as recycling reservoirs for some substances and as waste storage areas for others.

Color the ribosomes (M) and scattered patches of endoplasmic reticulum (N).

Ribosomes, though not membrane-bound, are usually considered to be organelles. *Ribosomes* function in structural protein and enzyme formation and may be found scattered in the cytoplasm and attached to various membranes. *Ribosomes* are frequently found attached to the *endoplasmic reticulum (ER),* a pervasive membrane system that courses through the cytoplasm, and also to the outer surface of the *nucleus.* The *endoplasmic reticulum* functions in the production of proteins and other substances and may form a continuous channel between the *nucleus* and *plasma membrane.*

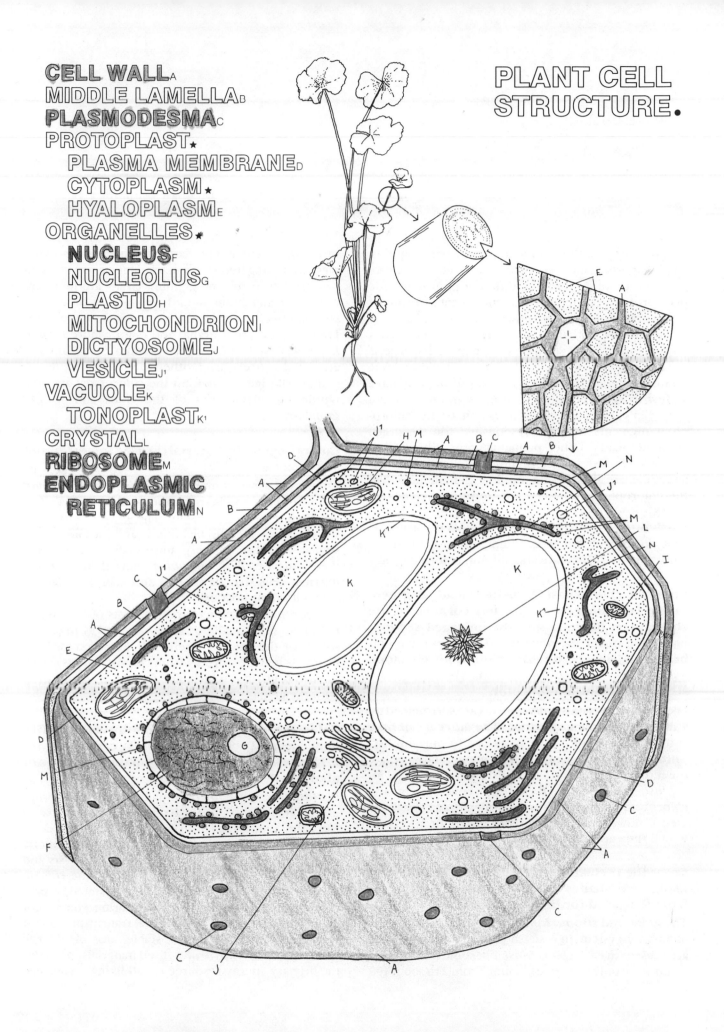

PLANT CELL STRUCTURE.

CELL WALL A
MIDDLE LAMELLA B
PLASMODESMA C
PROTOPLAST ★
  PLASMA MEMBRANE D
  CYTOPLASM ★
    HYALOPLASM E
ORGANELLES ★
  NUCLEUS F
  NUCLEOLUS G
  PLASTID H
  MITOCHONDRION I
  DICTYOSOME J
  VESICLE J¹
VACUOLE K
  TONOPLAST K¹
CRYSTAL L
RIBOSOME M
ENDOPLASMIC
  RETICULUM N

4
GREEN
SOLAR CELLS

Chloroplasts are the dominant plastid type found in all plant parts that are green, but the details of *chloroplast* structure vary among different photosynthetic plant groups. A generalized vascular plant *chloroplast* is illustrated. In photosynthetic vascular plants, or most land plants, *chloroplasts* are usually numerous within each photosynthetic cell, and their shape is usually ellipsoid to reniform (kidney-shaped). Some photosynthetic plants, such as many algae, may have as few as one *chloroplast,* often with an elaborate shape, in each cell. *Chloroplasts* contain the chlorophyll pigments that are responsible for the green color of plants. Other pigments, primarily xanthophylls (yellows) and carotenoids (yellows, oranges and reds), are also present, but in lesser amounts than the chlorophylls. The combination of different pigments produces various shades of green. Chlorophylls often effectively mask these other pigments except during autumn in deciduous plants, when the chlorophylls degenerate and autumn colors appear.

Color the chloroplast (A), using a shade of green, in the cell diagram (upper left corner). In the chloroplast diagram and the enlarged section, color the outer membrane (A¹) and inner membrane (A²) of the double membrane envelope using the same shade of green. Also, color the stroma (B) using a light color.

Like all other plastids, *chloroplasts* are surrounded by a double membrane envelope. The *outer membrane* and *inner membrane* are appressed to one another. Through selective control, these membranes mediate the movement of substances into and out of the *chloroplast.* The double membrane envelope of the *chloroplast* encloses a complex membrane system as well as other structures and the *stroma,* or matrix, in which they are suspended.

Color the thylakoids (C) and stroma lamellae (D) using two shades of green that are different from that used for the chloroplast.

The grana and *stroma lamellae* form an intrastromal membrane system. In vascular plants, numerous grana are present in each *chloroplast,* but some nonvascular plants have only one large granum. From 2 to about 100 *thylakoids,* which are flattened, membrane-bound sacs, are aggregated into a stack, similar to a stack of pancakes, to form one granum. Grana are interconnected by membrane-bound *stroma lamellae.* Each *stroma lamellum* spans the *stroma* between two *thylakoids* in different grana, and one or more *stroma lamellae* originate from each granum. Some lamellae are tubular (frets) and others are flattened (intergrana). All pigments within the *chloroplast* are held within the membranes of the grana and *stroma lamellae.*

Color the genetic material (E), ribosomes (F), and starch grains.

Chloroplasts also contain their own *genetic material* and *ribosomes* and increase their numbers by a process similar to cell division. The *genetic material,* which is not membrane-bound, and *ribosomes* are similar to those of very unspecialized cells. One theory suggests that *chloroplasts* have their origins as invasive cells that developed an exclusive relationship with their host cell.

Chloroplasts produce large amounts of food materials, but only a small portion is required as an energy source by the photosynthetic cell. Some excess food materials may be stored within *chloroplasts* as starch in *starch grains,* which are formed in association with the grana.

Color the chloroplast (A), photosynthesis (H), light energy (I), carbon dioxide (J), carbohydrates (K), water (L), and oxygen (M) in the schematic of photosynthesis at the bottom of the plate.

Photosynthesis may be briefly defined as the capture of *light energy* through the fixation of *carbon dioxide* from the atmosphere into energy-rich carbon compounds, or *carbohydrates.* In most photosynthetic organisms, *water* is required for *photosynthesis* and *oxygen* is liberated. The importance of the *chloroplast* in *photosynthesis* cannot be overstated; it provides the primary means of capturing solar *light energy* in living organisms. The transformation of unstable *light energy* into a stable, storable energy form held in *carbohydrates,* food materials, provides the primary energy resource for all living organisms.

CHLOROPLAST STRUCTURE AND FUNCTION.

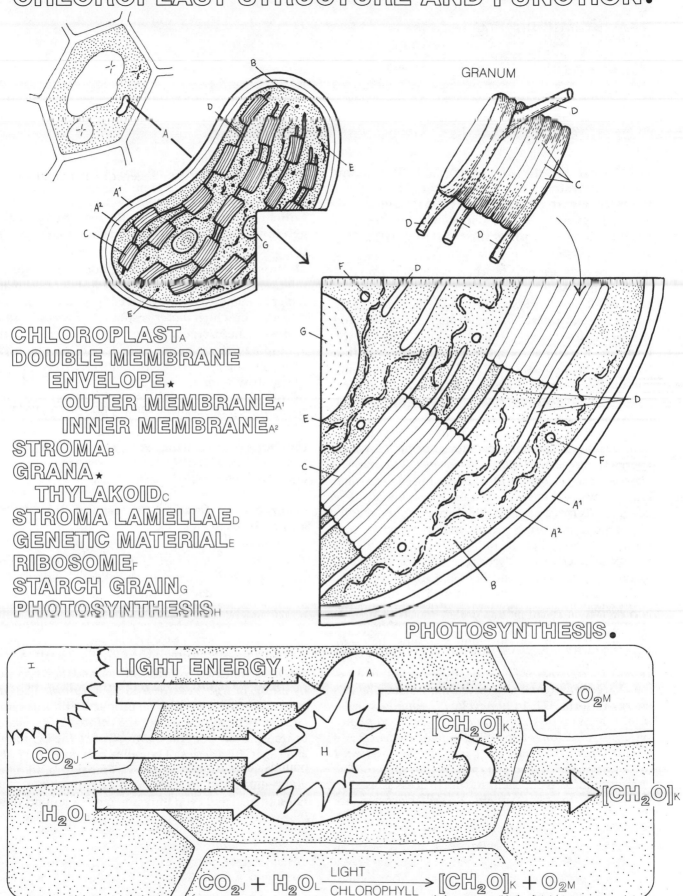

GRANUM

CHLOROPLAST$_A$
DOUBLE MEMBRANE
ENVELOPE ★
   OUTER MEMBRANE$_{A^1}$
   INNER MEMBRANE$_{A^2}$
STROMA$_B$
GRANA ★
   THYLAKOID$_C$
STROMA LAMELLAE$_D$
GENETIC MATERIAL$_E$
RIBOSOME$_F$
STARCH GRAIN$_G$
PHOTOSYNTHESIS$_H$

PHOTOSYNTHESIS.

LIGHT ENERGY

O_{2M}

$[CH_2O]_K$

CO_{2J}

H_2O_L

$[CH_2O]_K$

$$CO_{2J} + H_2O_L \xrightarrow[\text{CHLOROPHYLL}]{\text{LIGHT}} [CH_2O]_K + O_{2M}$$

5
CAPTURING SOLAR ENERGY

Color light energy (A), chloroplast (B), granum (C), photosystem I (E), photosystem II (E¹), electrons (F), electron excitation (G), ADP and Pi (H), ATP (I), cyclic photophosphorylation (J), and noncyclic photophosphorylation (J¹). Use shades of green for (C), (E) and (E¹).

In photosynthesis, which occurs within *chloroplasts* in most plants, *light energy* is transmitted through plant tissues to *chloroplast*-bearing cells. *Light energy* enters the *chloroplast,* passes through the stroma, and strikes the *light energy* collecting membranes of the *granum.* The initial receptors of *light energy* within the *granum, photosystem I* and *photosystem II,* are antennalike nets of linked chlorophyll molecules. Other *light energy* absorbing pigments may also be present.

Within the *granum,* a series of *light energy* dependent reactions, or light reactions, occur in which energy is transferred from *light energy* to *electrons.* In *electron excitation, electrons* are raised to a high energy level, or charged, by absorbing energy. The charged *electrons* are unstable, however, and readily discharge, by releasing energy, and drop back to their normal energy level. In photophosphorylation, the energy released by the discharging *electrons* is used to form an energy-rich bond between *ADP* (adenosine diphosphate) and *Pi* (phosphate). This forms *ATP* (adenosine triphosphate), which is the major energy-carrying compound in cells. Two types of photophosphorylation occur in the light reactions. In *cyclic photophosphorylation,* which involves only *photosystem I, electrons* are continually recycled, first being charged by *light energy* and then discharging this energy to form *ATP.* In *noncyclic photophosphorylation,* which involves both *photosystem I* and *photosystem II, electrons* are charged in *photosystem II* and then discharged as they are passed to *photosystem I* through a series of electron carriers (not shown). The energy drained from them in this process is used to form *ATP.* In *photosystem I,* the *electrons* are

recharged to a high energy level by *light energy* and then removed from the photosystem.

Color photolysis of water (K), water (L), hydrogen ions (M), oxygen (N), Hill reaction (O), NADP (P), and NADPH (Q).

Electrons removed from the photosystem during *noncyclic photophosphorylation* must be continually replenished by *electrons* derived from the *photolysis of water,* in which *water* is split into *hydrogen ions,* electrons, and *oxygen.* The *oxygen* is liberated to the atmosphere. A reaction coupled with the *photolysis of water,* the *Hill reaction,* pulls *electrons* through the photosystems and accepts *hydrogen ions* released by the *photolysis of water.* An electron acceptor, *nicotinamide adenine dinucleotide phosphate (NADP),* combines with the charged *electrons* from the photosystem and the *hydrogen ions* from *water* to form reduced *NADPH,* which is a second energy-rich compound and energy carrier.

Color carbon fixation (R), carbon dioxide (S), and carbohydrates (T).

The two energy carriers, *ATP* and *NADPH,* obtained from *light reactions* are transferred to the stroma, where the *light energy* independent, or dark reactions, occur. In dark reactions, *carbon fixation* links carbon from atmospheric *carbon dioxide* into energy-rich *carbohydrate* compounds by utilizing the energy held by *ATP* and *NADPH,* as well as the *hydrogen ions* from *NADPH.* In the process, *ADP, Pi* and *NADP* are released from the stroma and then recycled by means of *light reactions* in the *granum.* The *carbohydrate* formed is the end product of photosynthesis. It may be stored within the *chloroplast* as starch or transported from the *chloroplast* to the cytoplasm as a sugar. By a series of complex reactions, *carbohydrate* may be converted into various sugars, starches, and cellulose (all are carbohydrate compounds); into fats and oils; and even into amino acids and proteins.

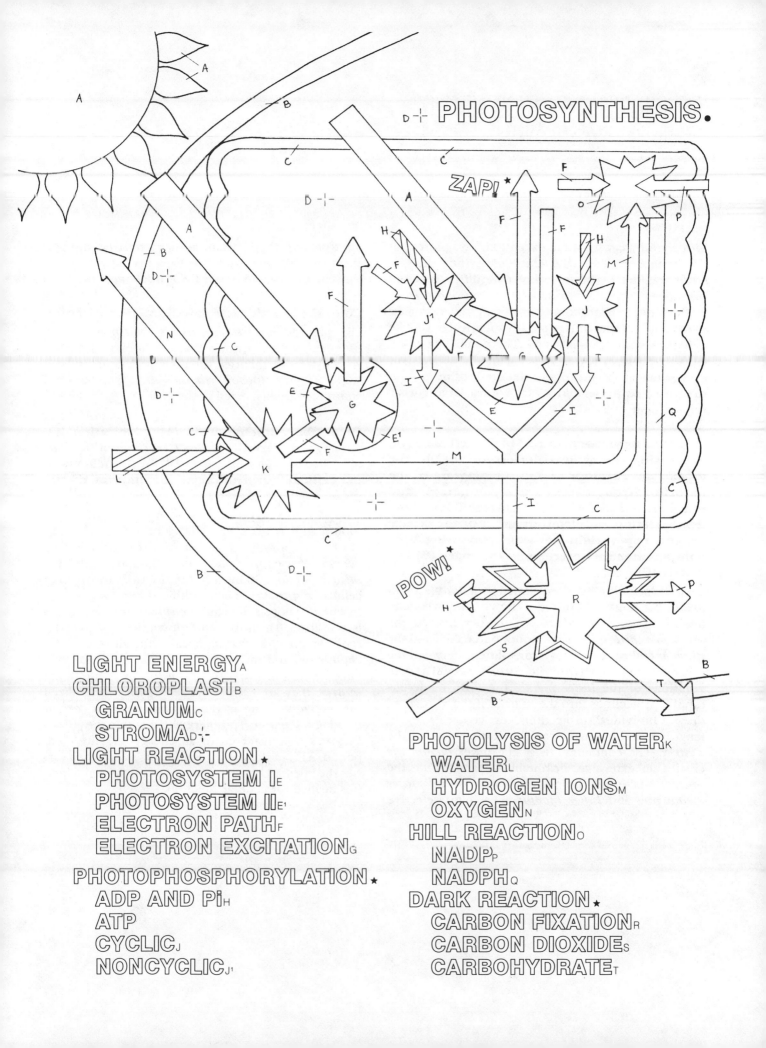

PHOTOSYNTHESIS.

ZAP! ★

POW! ★

LIGHT ENERGY A
CHLOROPLAST B
GRANUM C
STROMA D
LIGHT REACTION ★
PHOTOSYSTEM I E
PHOTOSYSTEM II E¹
ELECTRON PATH F
ELECTRON EXCITATION G

PHOTOPHOSPHORYLATION ★
ADP AND Pi H
ATP I
CYCLIC J
NONCYCLIC J¹

PHOTOLYSIS OF WATER K
WATER L
HYDROGEN IONS M
OXYGEN N
HILL REACTION O
NADP P
NADPH Q
DARK REACTION ★
CARBON FIXATION R
CARBON DIOXIDE S
CARBOHYDRATE T

6
CELLULAR ENERGY GENERATION

Mitochondria, with some exceptions, are present in all living cells. Like the chloroplast (not shown), structural details of *mitochondria* differ between organisms. Therefore, a generalized *mitochondrion* is illustrated. *Mitochondria* are small, much smaller than chloroplasts, and are usually spherical or rodlike cylinders with rounded ends. The number of *mitochondria* within a cell varies with the activity, or energy demands, of the cell. Hundreds of *mitochondria* may be present in a cell with a large energy requirement.

Color the mitochondrion (A) in the cell diagram, and, using the same color, color the double membrane envelope (A¹) in the diagram of the mitochondrion and the outer membrane (A²) and inner membrane (A³) in the enlarged mitochondrion section and detail diagram of the cristae (in the upper right corner). Also using light colors, color the intermembrane space (B) and matrix (C).

Each *mitochondrion* is enveloped by a *double membrane,* but unlike plastids (not shown), the two membranes are not appressed to one another. Instead, the *outer membrane* forms a smooth outer shell and the *inner membrane* is highly convoluted. Between the two membranes is an *intermembrane space.* The convolutions of the *inner membrane* form numerous tubular projections into the *matrix* of the *mitochondrion.* Individual projections, or cristae, vary in length, and some may span the entire width of the *matrix.* The configuration of the *inner membrane* creates an extensive membrane surface within the *matrix.* Like all living membranes, the *outer membrane* and *inner membrane* exhibit selective

control over the substances that may pass through them. In this way, they mediate the movement of substances into and out of the *mitochondrion.*

Color the genetic material (D) and ribosomes (E).

Like plastids, *mitochondria* have their own *genetic material,* which is not membrane-bound, and *ribosomes.* Similar theories for their origins have been suggested. In *mitochondria,* the *genetic material* and *ribosomes* are dispersed in the *matrix.*

Color the mitochondrion (A), respiration (F), energy (G), carbohydrates (H), oxygen (I), carbon dioxide (J), and water (K) in the schematic on cellular respiration at the bottom of the plate.

Mitochondria play a major role in the breakdown of food materials to release *energy* needed by the cell. Most of the enzymes active in cellular *respiration,* or *energy* release, are concentrated in and around the *mitochondria,* and most of the reactions involved in cellular *respiration* occur within the *mitochondria.* In a brief synopsis of cellular *respiration,* the process is as follows. The initial breakdown of the food materials, or *carbohydrates,* occurs outside the *mitochondrion,* and the smaller carbon compounds produced enter the *matrix* of the *mitochondrion.* In the *matrix,* they are further broken down and oxidized by *oxygen* with the accompanying release of *carbon dioxide* and *water.* The energy held within the *carbohydrates* is transferred to an energy-carrying compound, ATP, in the *inner membrane,* which then carries the released *energy* to areas within the cell where it is needed for cellular function.

MITOCHONDRION STRUCTURE AND FUNCTION.

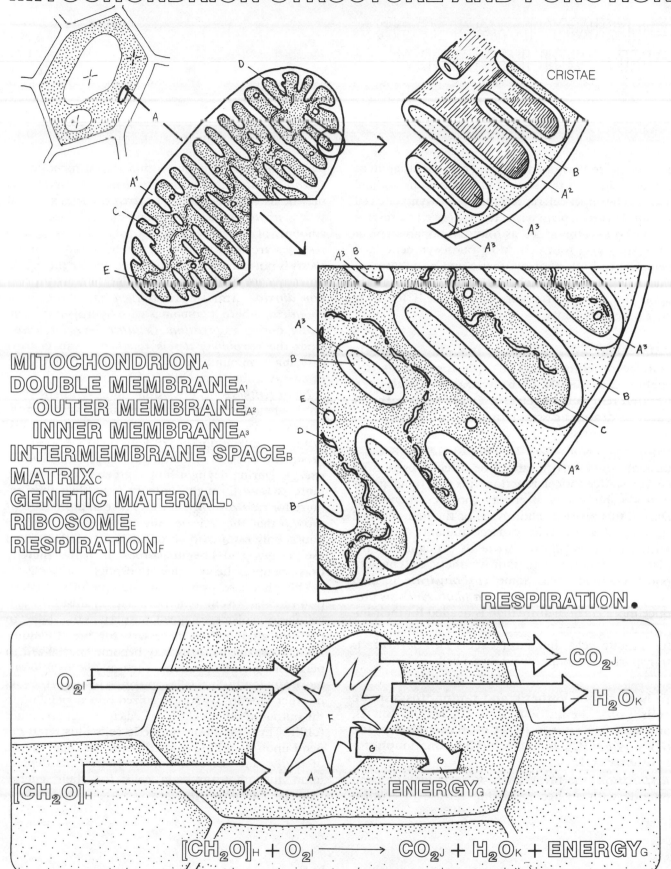

CRISTAE

MITOCHONDRION$_A$
DOUBLE MEMBRANE$_{A^1}$
  OUTER MEMBRANE$_{A^2}$
  INNER MEMBRANE$_{A^3}$
INTERMEMBRANE SPACE$_B$
MATRIX$_C$
GENETIC MATERIAL$_D$
RIBOSOME$_E$
RESPIRATION$_F$

RESPIRATION.

O_{2I}

CO_{2J}

H_2O_K

$[CH_2O]_H$

ENERGY$_G$

$$[CH_2O]_H + O_{2I} \longrightarrow CO_{2J} + H_2O_K + ENERGY_G$$

7
ENERGY TRANSFER AND UTILIZATION

Two cells are illustrated in this schematic diagram of energy capture, transfer, and utilization in land plants. The upper cell represents a photosynthetic cell as found in green plant parts; the lower cell represents a nonphotosynthetic cell as found in nongreen plant parts such as a root cell. Most photosynthetic land plants have both photosynthetic and nonphotosynthetic cells, and the nonphotosynthetic cells must rely upon the output of photosynthetic cells for their energy requirements.

Color light energy (A). In the photosynthetic cell, color chloroplast (B), photosynthesis (C), carbon dioxide (D), carbohydrate (E), water (F), oxygen (G), and starch (H). Color total paths of carbohydrate and water in both cell types.

In photosynthetic cells, *light energy* is captured within the *chloroplasts* by *photosynthesis*. Because unstable *light energy* cannot be directly used as an energy source for cell function, it is first transformed into a stable chemical energy form, *carbohydrate*. During this process, photolysis, or the light-induced splitting of *water,* releases large amounts of *oxygen* within the *chloroplasts*. *Water* is brought into the plant and transported to photosynthetic cells by nonphotosynthetic cells. Some *carbohydrate* is used directly, but cells specialized for *photosynthesis* produce more *carbohydrate* than is needed for their immediate use. Some is stored as *starch* within the photosynthetic cells, but most excess *carbohydrate* is transported to nonphotosynthetic cells.

In both photosynthetic and nonphotosynthetic cells, color the mitochondria (I), respiration (J), and cellular energy (K). Color carbon dioxide (D), oxygen (G), and starch (H) in the nonphotosynthetic cell.

Respiration, the breakdown of *carbohydrate* compounds with an accompanying release of *cellular energy,* occurs primarily within the *mitochondria*. *Carbohydrates* formed directly from *photosynthesis* may be used by the cell, but *starch* reserves are used when *photosynthesis* is supplying an inadequate amount of *carbohydrate*. At night, only *starch* reserves are used in most plants because *photosynthesis* is not occurring. Carbon released by the breakdown of *carbohydrate* compounds is liberated as *carbon dioxide*. Atmospheric *oxygen* enters the *mitochondria,* where it combines with hydrogen to form *water* during *respiration*. *Cellular energy* released from the *carbohydrates* is retained in an energy-carrying compound, ATP, which distributes it to areas of energy utilization within the cell.

Photosynthesis occurs within photosynthetic cells only when *light energy* is available. *Respiration,* however, is a continuous process within both photosynthetic and nonphotosynthetic cells because life processes demand a continual supply of *cellular energy*. During daylight hours, green plants liberate more *oxygen* through *photosynthesis* than they require for *respiration*. They also take in more *carbon dioxide* than they release. However, during nighttime hours, only *respiration* is occurring, and plants then use *oxygen* and liberate *carbon dioxide,* thereby drawing upon the available atmospheric *oxygen*.

This plate also presents an example of specialization of cells within a multicellular organism. In unicellular organisms, each cell must retain the ability to perform all functions necessary for life. In multicellular organisms, cells may become specialized to perform specific functions, such as *photosynthesis,* more efficiently. Specialization often involves the loss of other functions, and specialized cells depend upon the output of other cells for needs they can no longer fulfill. Thus, the photosynthetic cells illustrated depend upon nonphotosynthetic cells for their supply of *water,* and the nonphotosynthetic cells depend upon the photosynthetic cells for their energy material, *carbohydrate*.

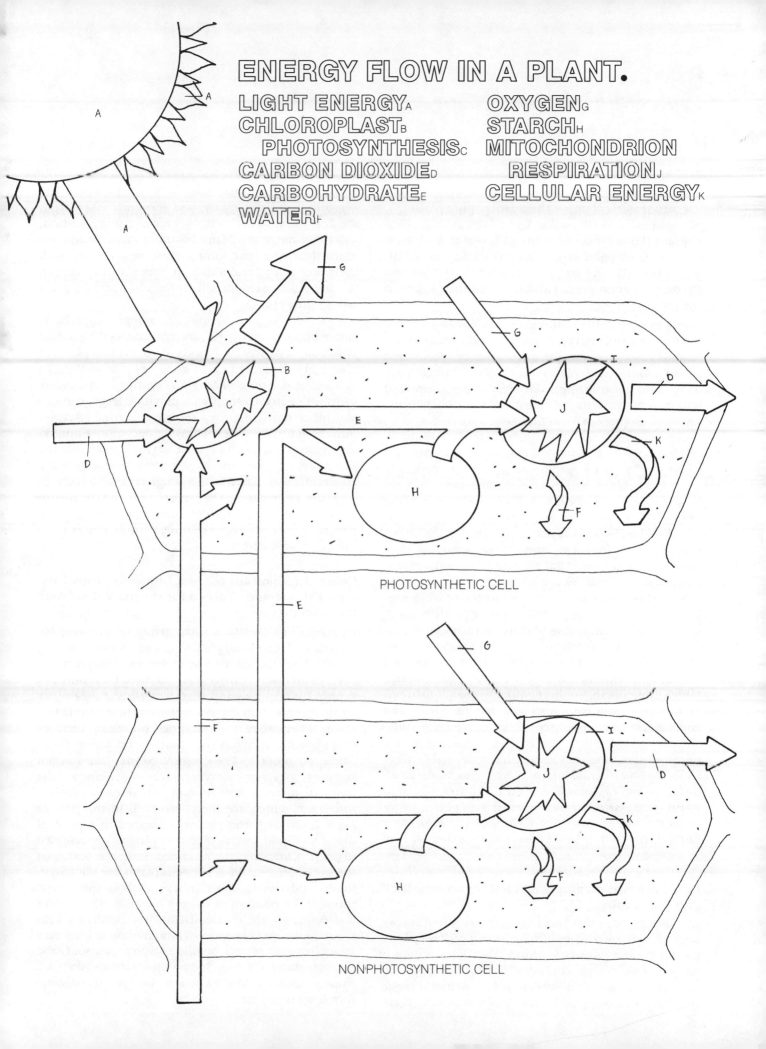

ENERGY FLOW IN A PLANT.

LIGHT ENERGY A   OXYGEN G
CHLOROPLAST B   STARCH H
    PHOTOSYNTHESIS C   MITOCHONDRION I
CARBON DIOXIDE D   RESPIRATION J
CARBOHYDRATE E   CELLULAR ENERGY K
WATER F

PHOTOSYNTHETIC CELL

NON-PHOTOSYNTHETIC CELL

8
MODES OF PLANT NUTRITION

For the diagrams labeled "autotrophic" and "semiparasitic," color the autotrophic plant (A), photosynthesis (B), solar light energy (C), chloroplast (D), carbon dioxide (E), water and minerals (F), oxygen (G), semiparasitic plant (H), host plant (I), and organic nutrients (J). Use two shades of green for (A) and (H), and the (A) green for (I).

Plants may be divided into groups according to how they obtain the energy resources and organic compounds they require for their energy needs and growth. Only green plants are capable, through *photosynthesis,* of utilizing *solar light energy,* captured within *chloroplasts* in most, for manufacturing organic compounds from inorganic substances.

Most green plants, from unicellular algae to multicellular flowering plants, are entirely *autotrophic,* requiring only an inorganic energy source, primarily *solar light energy,* and inorganic substances, primarily *carbon dioxide, water,* and *minerals,* to form the organic compounds necessary for life. *Carbon dioxide,* absorbed by aquatic plants from water and by terrestrial plants from the atmosphere, serves as their only carbon source. Like all photosynthetic plants, *autotrophic plants* release *oxygen.*

Not all green plants are entirely *autotrophic.* Some, such as *semiparasitic plants,* are only partially *autotrophic* and are incapable of manufacturing all the organic compounds they need from inorganic substances. *Semiparasitic* plants require a direct physical connection with an *autotrophic host plant* through which they obtain some required *organic nutrients* as an energy and carbon source, but they have two energy and carbon sources. Like completely *autotrophic plants,* they have green parts and are photosynthetic, so they also utilize *solar light energy* to fix *carbon dioxide* into organic compounds. In addition to the *organic nutrients* taken from the *host plant, semiparasites* usually must also obtain *water* and *minerals* through the tissues of the *host plant* because they often lack a direct connection with an inorganic source.

Color the two diagrams labeled "holoparasitic" and "saprophytic."

Holoparasitic plants also have a direct physical connection with the tissues of an *autotrophic host plant,* but unlike *semiparasitic plants,* they have no green parts and are not photosynthetic. Therefore, they do not utilize *solar light energy* to fix *carbon dioxide* into organic compounds and do not release *oxygen.*

Holoparasitic plants must rely solely upon the *organic nutrients, water,* and *minerals* taken from the *host plant* as their only source of energy, carbon, and other nutrients. Many *holoparasites* that live for more than one year form a close relationship with their *host plant,* and although they rob and weaken the *host,* they seldom kill it, for the parasitic plant would then also die.

Saprophytic plants, though nonphotosynthetic and without green parts, are not attached to a *host plant. Saprophytes* are found growing as independent plants in substrates rich in *dead organic matter.* In order to utilize the *dead organic matter* as an *organic nutrient* source to supply them with energy, carbon, and other nutrients, they secrete *digestive enzymes* into the substrate. These *enzymes* decompose nonliving organic matter and release *organic nutrients* the saprophyte can absorb. Most *saprophytic plants* are apparently associated with a *fungus* that also secretes *digestive enzymes* into the substrate to aid the *saprophyte* in obtaining *organic nutrients.* In return, the *fungus* apparently receives some carbohydrates from the *saprophytic plant.*

Color the diagram labeled "insectivorous" using a third shade of green for the plant (P). Color the insects (Q).

Insectivorous plants, a small group of *autotrophic* aquatic and terrestrial plants typically found growing in substrates lacking or low in several mineral nutrients, particularly nitrogen, supplement their nutrient supply by trapping and digesting *insects* and other small animals. Numerous adaptations lure *insects* to a plant, where they are entrapped on sticky surfaces (not shown), in snares (not shown), or within fluid-filled chambers. The *insectivorous plant* then secretes *digestive enzymes* that decompose the trapped insect's soft tissues to release *organic nutrients,* which are then absorbed. In the pitcher plant illustrated, the victim is trapped in a pool of digestive liquid, and as digestion progresses, released *organic nutrients* are absorbed into the leaf that forms the pitcher. *Insectivorous* plants are photosynthetic and obtain most of the carbon they need through the fixation of *carbon dioxide.* Also, *water* and *minerals* are obtained from the nonliving substrate in which they grow. Though in most instances *insectivorous plants* would survive without the *organic nutrients* supplied by their victims, they are usually more robust when a supply of *organic nutrients* is available.

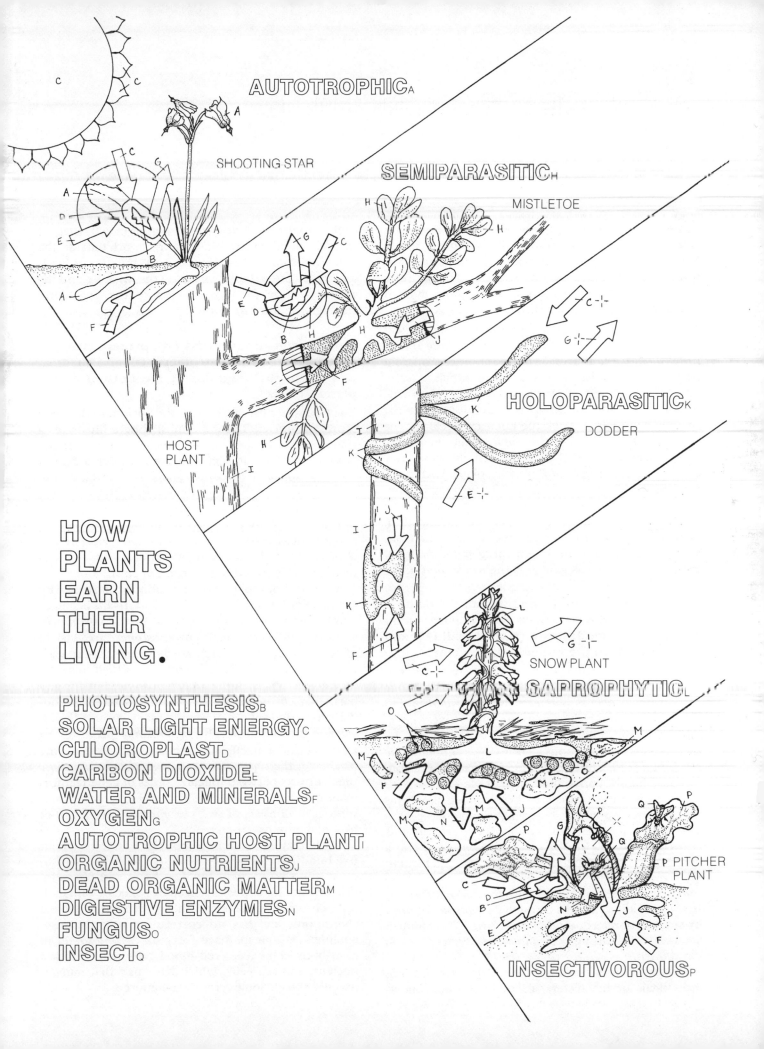

AUTOTROPHIC A

SHOOTING STAR

SEMIPARASITIC H

MISTLETOE

HOST
PLANT

HOLOPARASITIC K

DODDER

HOW
PLANTS
EARN
THEIR
LIVING.

SNOW PLANT

SAPROPHYTIC L

PHOTOSYNTHESIS B
SOLAR LIGHT ENERGY C
CHLOROPLAST D
CARBON DIOXIDE E
WATER AND MINERALS F
OXYGEN G
AUTOTROPHIC HOST PLANT I
ORGANIC NUTRIENTS J
DEAD ORGANIC MATTER M
DIGESTIVE ENZYMES N
FUNGUS O
INSECT Q

PITCHER
PLANT

INSECTIVOROUS P

9
THE NUCLEAR CONNECTION

Color the nucleus (A) in the small cell diagram. In the boxed enlargement, color the outer membrane (A¹), inner membrane (A²), and nuclear pores (B) of the nuclear envelope and the genetic material (C), nucleolus (D), and, using a light color, the nucleoplasm (E).

The *nucleus* is bound by a double-membrane nuclear envelope that consists of an *outer membrane* and an *inner membrane* appressed to one another. Apparent perforations, *nuclear pores,* are thin circular (octagonal) discs in the nuclear envelope rather than true perforations. *Nuclear pores* may function in moving specific materials through the nuclear envelope. The *genetic material* within the *nucleus* determines the form and function of the mature cell and ultimately the form and function of the entire plant. Within the *nucleus,* the *nucleolus* functions in forming ribonucleic acid (RNA), an important component in the mechanism for interpreting genetic information. Some plants have *nuclei* with two or more *nucleoli.* The matrix material in which the *genetic material* and *nucleolus* are suspended is the *nucleoplasm.*

Color the cytoplasm (F), endoplasmic reticulum (G), intermembrane space (H), plasma membrane (I), plasmodesmata (J), cell wall (K), and ribosomes (L) in the boxed enlargement.

The *outer membrane* of the nuclear envelope is in contact with the *cytoplasm,* and a selective exchange of substances occurs between the *nucleoplasm* and *cytoplasm* through the semipermeable nuclear envelope. At some points, the *outer membrane* can be seen to be continuous with the *endoplasmic reticulum (ER).* The *endoplasmic reticulum* is a continuous, highly convoluted membrane system with varying amounts of *intermembrane space* that forms sheets, tubular pockets, and sacs throughout the *cytoplasm.* In some areas, the *endoplasmic reticulum* is also continuous with the *plasma membrane,* and a continuous channel between the *nucleus* and *plasma membrane* may be formed. The *endoplasmic reticulum* may also be continuous with the *endoplasmic reticulum* of adjacent cells through the *plasmodesmata* by strands, or microtubules (not shown), that connect adjacent cells through the *plasmodesmata* pores in the *cell wall.*

Ribosomes may be found free in the *cytoplasm* as individuals or in clusters, called polysomes, or at-

tached to the *endoplasmic reticulum* in cells that are actively building proteins for transport outside the cell. Ribosomal RNA produced by the *nucleolus* is assembled into *ribosomes* within the *nucleus,* and the *ribosomes* are then transferred to the *cytoplasm* through the *nuclear pores.*

Color chromatin (C), DNA (M), protein coat (N), chromosome (C¹), centromere (O), and chromatids (P) in the diagrams at the bottom of the plate.

The two spherical diagrams at the bottom of the plate represent *nuclei.* The uppermost sphere illustrates a nondividing, or interphase, *nucleus* in which the *genetic material* is highly diffuse. Under a microscope, it appears as a dense, granular material within the *nucleoplasm,* but it is illustrated as strandlike for coloring purposes. *Genetic material,* known as *chromatin* when fully extended in the nondividing cell, consists of two parallel strands of *deoxyribonucleic acid (DNA)* twisted about one another in a helical manner and covered with a *protein coat.*

In a cell about to undergo division, as illustrated by the lowermost sphere, the *genetic material* becomes condensed by tight spiral coiling of the *chromatin* strands, and individual *chromosomes* are visible under a microscope. Each *chromosome* has at least one constriction, the *centromere,* which may be located anywhere along a *chromosome;* but for any particular *chromosome,* its position is fixed. Secondary constrictions are present in many *chromosomes.*

Prior to division, each *chromosome* produces an exact copy of itself that remains attached to the original at the *centromere,* and a single *chromosome* then consists of two duplicate copies of genetic information, called *chromatids.* Therefore, each *chromosome* now consists of two *chromatids* held together by a common *centromere.*

There may be from two (rare) to hundreds (about five hundred) *chromosomes* present in each cell of an organism. *Chromosome* number is constant for an individual and for a species, except for species with multiple chromosome sets and those with chromosomal aberrations, and has no correlation with organism complexity. Humans have forty-six *chromosomes* in each body cell except red blood cells, which lack a nucleus, and sex cells, which have half that number (twenty-three). Some ferns have hundreds.

THE NUCLEUS AND ASSOCIATED STRUCTURES.

NUCLEUS$_A$
   NUCLEAR ENVELOPE★
     OUTER MEMBRANE$_{A^1}$
     INNER MEMBRANE$_{A^2}$
     NUCLEAR PORE$_B$
   GENETIC MATERIAL$_C$
   NUCLEOLUS$_D$
   NUCLEOPLASM$_E$
CYTOPLASM$_F$
ENDOPLASMIC
   RETICULUM$_G$
     INTERMEMBRANE
       SPACE$_H$
PLASMA MEMBRANE$_I$
PLASMODESMA$_J$
CELL WALL$_K$
RIBOSOME$_L$

GENETIC MATERIAL★
   CHROMATIN$_C$
     DNA$_M$
     PROTEIN COAT$_N$
   CHROMOSOME$_{C^1}$
     CENTROMERE$_O$
     CHROMATID$_P$

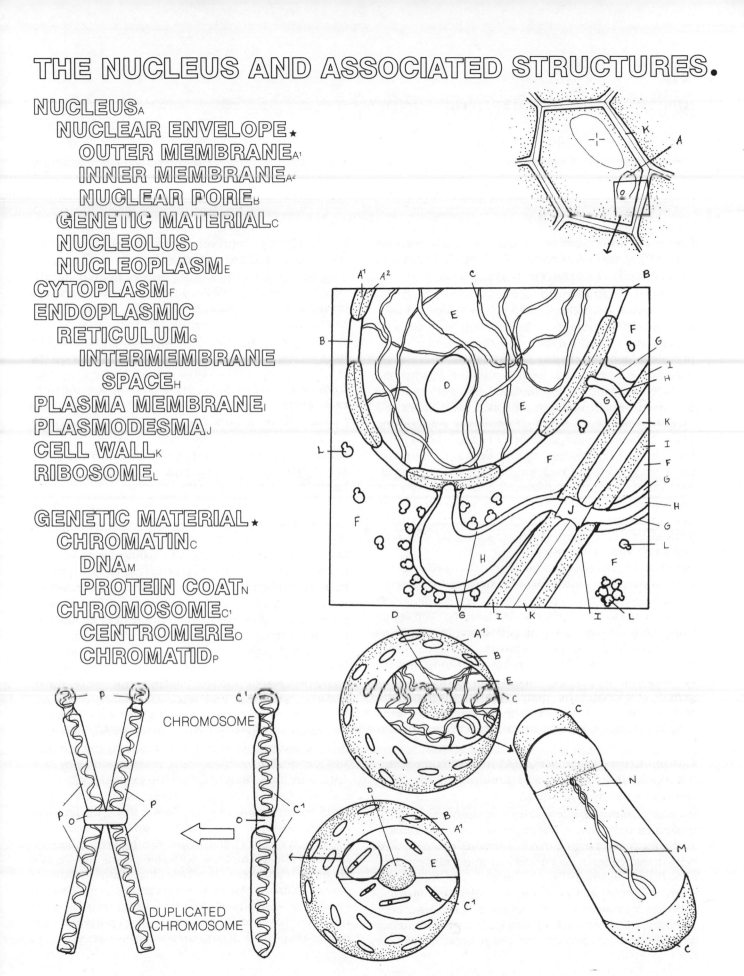

CHROMOSOME

DUPLICATED CHROMOSOME

10
MITOTIC CHROMOSOME BEHAVIOR

Color chromosome set A (A) and chromosome set B (B) at early interphase for both haploid and diploid cells. Use two contrasting colors.

Cellular division that produces two new cells genetically identical to the parent cell is called mitosis. Each new cell contains a chromosome set that is an exact copy of the parent cell's chromosome set. It produces genetic rigidity, or stability, because no variation is introduced. Both haploid cells, with one set of chromosomes, and diploid cells, with two sets of chromosomes, undergo mitosis. This is illustrated by *chromosome set A* and *chromosome set B* on the plate. In haploid cells, only one of each kind of chromosome is present, but diploid cells have homologous pairs, or two of each kind of chromosome, present. Each member of a homologous pair has a sequence of genes that matches the sequence on the other member. The two homologous pairs in the diploid cell illustrated are represented by two long and two short chromosomes.

Prior to division, a cell must duplicate its chromosomes in order for each new cell to receive identical sets. This duplication occurs during interphase, a period of high cellular activity. The interphase cells illustrated are in early interphase prior to chromosome duplication. Though individual chromosomes are illustrated, interphase chromosomes are actually diffused as chromatin and functioning as genetic material. The plate's highly diagrammatic aspect is intended to clarify the chromosomal events of mitosis rather than to depict actual appearance.

Color chromosome set A (A) and chromosome set B (B) at prophase and metaphase for both haploid and diploid cells.

As a cell enters prophase, the now duplicated chromosomes condense to become shorter and thicker and appear as a tangled mass of strands when viewed with a microscope. As condensation continues, individual chromosomes become evident. Each chromosome appears as two strands, called chromatids, which are attached at a common point, the centromere. At the end of prophase, the chromosomes are scattered in the cell, and the nuclear membrane and nucleolus, represented by the dotted circles, break down and are no longer present.

In metaphase, the chromosomes migrate to the center of the cell and become individually arranged on the equatorial plate, a positional and not a structural feature that is perpendicular to the direction of division. A mitotic spindle apparatus, represented by lines on the metaphase diagrams, develops across the cell in the direction of division. A spindle fiber runs from each end, or pole, of the spindle apparatus to each centromere. Therefore, each centromere has two spindle fibers, one from each pole, attached to it.

Color chromosome set A (A) and chromosome set B (B) at anaphase and telophase for both haploid and diploid cells.

In anaphase, the spindle apparatus appears to function in separating the two chromatids into two individual chromosomes. By this process, called karyokinesis, two sets of identical chromosomes are produced whether the dividing cell is haploid or diploid. Each new chromosome then begins movement toward its respective spindle pole in a manner that suggests it is being pulled by the spindle fiber attached to its centromere.

Telophase begins as the two sets of identical chromosomes reach the vicinity of their poles. As telophase progresses, the spindle apparatus breaks down and the nucleolus and nuclear membrane reform. Upon separation by a new cell wall, or cytokinesis, the two new cells reenter interphase. Each can be seen to have a chromosomal set identical to the parent's and identical to each other's. Cytokinesis may or may not occur immediately following karyokinesis.

Growth due to cell enlargement is apparently limited by the amount of cytoplasm a single nucleus can control. Mitosis provides an additional basis for growth in multicellular organisms by increasing cell number. For example, a pine tree begins its life as a single cell, but through mitosis, it contains billions of cells at maturity. Mitosis also functions in asexual, or vegetative, reproduction by producing single cells or multicellular units that have the potential for developing into a new generation of progeny.

ASEXUAL CELL DIVISION.

CHROMOSOME SET A_A
CHROMOSOME SET B_B

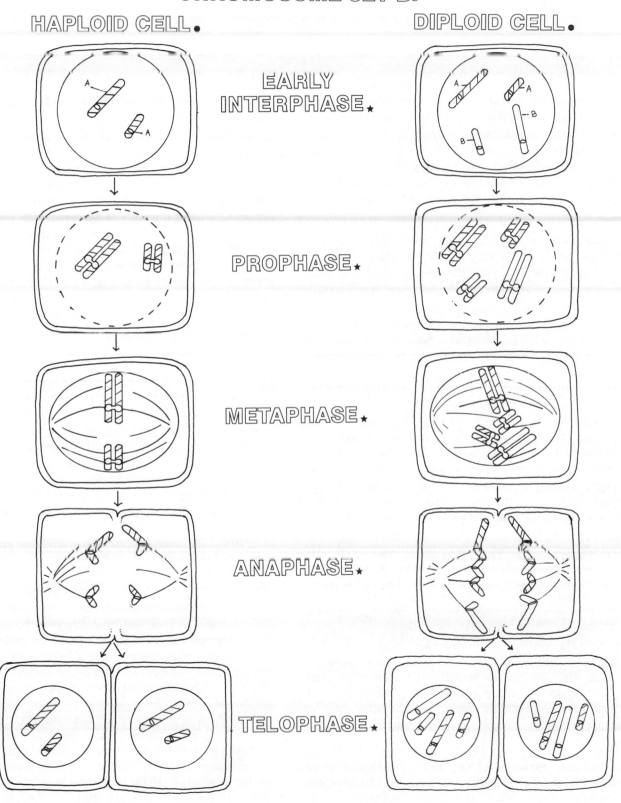

HAPLOID CELL.   DIPLOID CELL.

EARLY INTERPHASE.

PROPHASE.

METAPHASE.

ANAPHASE.

TELOPHASE.

11

SIMPLE GROWTH

The fate of cells produced by cell division is varied. Some enter the differentiation and maturation process to become specialized cell types no longer capable of cell division; others retain their capacity to divide. The planes of cell division and whether or not the cells produced remain attached to one another have a major influence on the structure of the mature plant.

Color planes of cell division, with the parent cell (A), separate cells (B), one-dimension division with cohesion (C), two-dimension division with cohesion, (D) and three-dimension division with cohesion (E).

In unicellular plants, such as the green alga *Chlamydomonas,* after division of a *parent cell,* offspring cells *separate* from one another to function as independent, individual plants. Because there is no cohesion, fusion of like parts, between cells, only unicellular structures are produced.

Multicellular plants are formed if cohesion occurs between adjacent cells following *division.* Repeated *divisions* in only *one dimension* produce a long chain, or filament, of cells. This is typical of many fungi and algae, such as the green alga *Ulothrix.* Branching occurs when a cell in the filament divides at an angle to the filament axis. When *division* occurs in *two dimensions* with cohesion, a sheet of cells one cell thick is produced. A few marine algae that have this growth form, such as the green alga *Monostroma,* may become quite large, up to several square feet, yet remain only one cell thick. If *divisions* occur in *three dimensions* with cohesion between all adjacent cells, a cluster of cells is formed. *Ulva* (not shown), a green marine alga similar to *Monostroma,* exhibits this type of growth but is only two cell layers thick because a single *division* occurs in the third dimension. Random *divisions* in *three dimensions* produce aggregations of cells such as found in the fleshy part of an apple.

Color the parent cell (A), cells exhibiting division (F), the cell exhibiting growth (G), and the cells exhibiting both division and growth. Use light colors for (F) and (G).

Growth may be defined as an irreversible increase in size and may not include cell *division.* Cell *division* only, with no enlargement occurring, does not constitute *growth.* In some forms of asexual and sexual reproduction, a cell may undergo a series of mitotic *divisions* within the wall of a *parent cell.* With each *division,* individual cell size decreases and no growth has occurred, only cell *division.* In all multicellular organisms, *growth* is due to a combination of cell *division* and enlargement. Cell *division* provides new cells as a basis for growth, and cell enlargement provides for an increase in size. In combination, they constitute *growth.*

Color the filament exhibiting intercalary growth. Note that the parent cells (A) undergo division (F) and then cell growth (F + G). Now color the filament exhibiting apical growth.

In some plants, *growth* is localized in specific areas; in others, it occurs randomly throughout the plant. Many filamentous plants, such as some algae and fungi, exhibit random, or intercalary, *growth.* In many of these plants, any cell within the plant may undergo *division.* The two cells produced then undergo *growth* and may be active as a *parent cell* at a later time. The plant illustrated, *Ulothrix,* grows by intercalary growth.

Apical growth is due to *division* and *growth* of a cell localized at the tip, or apex, of a filament. The filamentous red alga illustrated has a single apical cell that produces new cells. Upon *division,* the proximal cell of the pair of new cells undergoes *growth* and then enters a maturation process. Although the new apical cell also undergoes *growth,* it retains its capacity to divide. The other cells of the filament do not divide.

Growth, or enlargement, of a cell may be due to an irreversible uptake of water, an increase in cytoplasmic content, or a combination of both. In many plants, the cell vacuoles pull in water, creating pressure against the cell wall to stretch it to mature size.

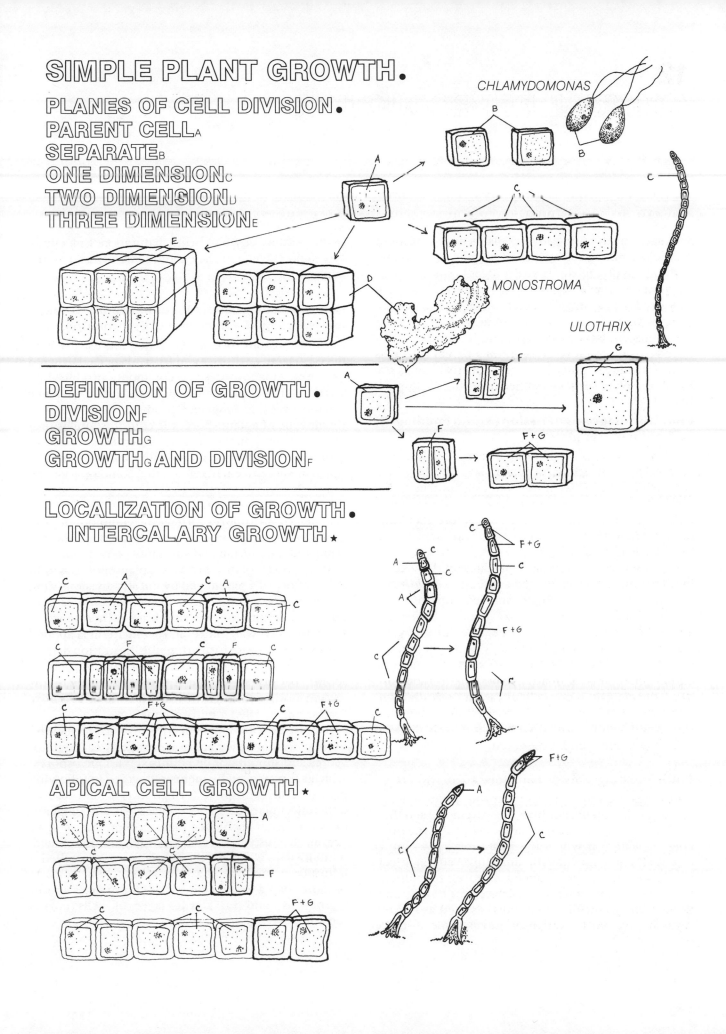

SIMPLE PLANT GROWTH.

PLANES OF CELL DIVISION.
PARENT CELL A
SEPARATE B
ONE DIMENSION C
TWO DIMENSION D
THREE DIMENSION E

CHLAMYDOMONAS

MONOSTROMA

ULOTHRIX

DEFINITION OF GROWTH.
DIVISION F
GROWTH G
GROWTH G AND DIVISION F

LOCALIZATION OF GROWTH.
INTERCALARY GROWTH ★

APICAL CELL GROWTH ★

12
ASEXUAL REPRODUCTION

Asexual reproduction provides plants with a means of producing new generations, or population increase, without involving the sexual process. Since mitosis produces asexually derived progeny, every individual produced by a single parent plant has a genetic composition identical to the parent. Both unicellular and multicellular asexual reproductive units are produced by one or more divisions of a parent cell. This plate introduces unicellular reproductive units; the following plate surveys multicellular units.

Color the diagrams of fission (A) and budding (B) at the top of the plate.

Fission is the division of a single parent cell into two new cells, or progeny, of equal size. The two new cells are produced separate from one another to become a new generation of independent individuals. New cells must attain mature size before they may divide. *Fission* is typical of bacteria and some unicellular algae and fungi.

A related type of asexual reproduction that occurs in some unicellular plants, such as yeast, is *budding*. A parent cell buds off a single, smaller, copy of itself, which then grows to mature size. Because *budding* is a slow process that cells may begin before they are mature, a short chain of cells may be formed in the process of *budding*. Unlike in *fission,* the budding parent cell retains its identity. In both *fission* and *budding,* the net gain is one cell for each mitotic division. In *fission,* two new cells are formed; in *budding,* a single new cell is formed with each mitotic division since the parent cell is conserved.

Color the diagrams of zoospore formation (C).

Zoospores are motile spores (capable of movement) produced within the confines of a parent cell and then released by rupture of the parent cell wall. The presence of water, in which the *zoospores* can swim, is essential, and many aquatic unicellular and multicellular algae and fungi produce *zoospores*. In multicellular plants that attach to a surface, *zoospores* function as important dispersal units, as well as asexual reproductive units, and may travel some distance

before settling down to begin maturation. Unicellular types may travel, but they mature directly into motile adults.

Color the diagrams of conidiospore (D) and mitosporangial aplanospore (E) formation.

Aplanospores are nonmotile. They are usually produced in large quantities and freely dispersed by wind or water. Aplanospores are commonly produced by terrestrial lower plants, the algae and fungi, but some aquatic algae and fungi also produce aplanospores.

One type of aplanospore, the *conidiospore,* is produced by some fungi, such as *Aspergillus. Conidiospores,* or conidia, are continually pinched from the tip of a specialized filament, the conidiophore, by mitosis in a process similar to *budding*. Since *conidiospores* are not released until they are mature, chains of beadlike *conidiospores* form at the conidiophore tip. *Conidiospores* are very light and account for the rapid and long-distance dispersal of some fungi.

Many lower plants produce aplanospores within the confines of a parent cell by mitotic divisions similar to *zoospore* formation, except the spores produced are nonmotile. Spore release is by rupture of the parent cell wall. Since no specific term is applied to this type of spore formation, the illustration depicting this process is labeled *"mitosporangial aplanospores"* because a parent cell within which spore formation occurs through mitosis, including *zoospore* formation, is called a mitosporangium.

Because asexually produced spores are formed by mitotic division, they are more properly called mitospores and should not be confused with meiospores, which are produced by meiosis (covered in following plates). True mitospore formation is widespread in both unicellular and multicellular algae and fungi but does not occur in higher plant groups. Any cell or organized cluster of cells in which mitosis occurs within each cell to produce spores may be called a mitosporangium. If the mitosporangium consists of more than a single cell, each cell produces spores by mitosis, and no sterile layer of cells covers the mitosporangium.

UNICELLULAR UNITS.

FISSIONₐ

BUDDINGʙ

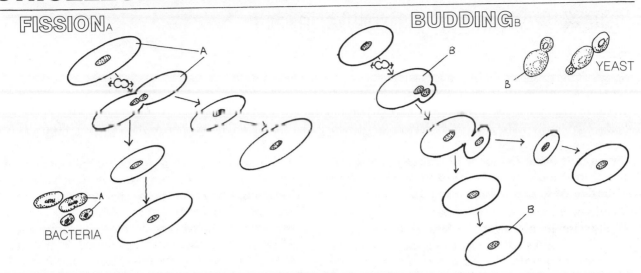

YEAST

BACTERIA

MITOSPORE FORMATION ★

ZOOSPORESᴄ

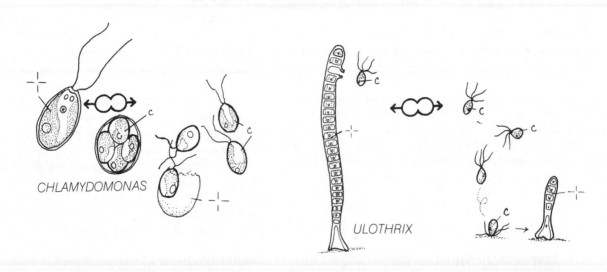

CHLAMYDOMONAS

ULOTHRIX

CONIDIOSPORESᴅ

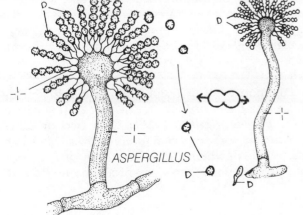

ASPERGILLUS

MITOSPORANGIAL APLANOSPORESᴇ

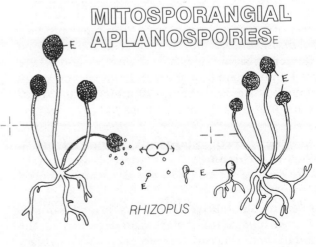

RHIZOPUS

13
ASEXUAL REPRODUCTION

Color parent plant (A), new plants (B), and gemmae (C), in the diagram labeled "gemmae." Use two shades of green for (A) and (B).

Multicellular asexual reproductive units range from relatively simple masses of a few cells to highly organized structures containing thousands of cells. As with the unicellular units, their function is to provide a means of population increase in which the *new plants* (progeny) are genetically identical to the *parent plant* and, in some, to provide a means of dispersal. Due to size, weight, and position of the reproductive units on the *parent plant*, dispersal is often limited.

Some plants, almost exclusively mosses and liverworts, produce *gemmae*, which are small, organized masses of few to numerous cells. *Gemmae* are formed externally and not within a parent cell. In some mosses, specialized portions of stems or leaves function as *gemmae*; in others, *gemmae* consist of distinct masses of cells produced by stems or leaves (not shown). Many liverworts produce *gemmae* within specialized cup-shaped structures called gemmae cups. Each *gemma* may produce a *new plant* under favorable conditions.

Color the parent plants (A) and the bulbils (D) in the diagram labeled "bulbils."

Bulbil, used as a general term, refers to any small, bulblike offset from a *parent plant*. *Bulbils* are typically swollen with reserve food materials and may be derived from modified roots or stems or compact clusters of modified leaves. Each *bulbil* may produce a *new plant*. *Bulbils* may be produced above ground on flowering stems in place of flowers and where leaves join the stem (not shown). Below ground, *bulbils* may be formed at the tips of underground stems or roots or at the base of a bulbous *parent plant*.

Color the parent plants (A), plantlets (B¹), stolons (E), rhizomes (F), roots (G), leaves (H), and flowering stems (I) in the diagrams labeled "plantlets."

Plantlets are diminuitive *new plants* formed as offsets from a *parent plant* that remain attached to the *parent plant* during early development. Various plant parts may give rise to *plantlets*. Some plants produce *plantlets* at intervals along *stolons*, which are slender, elongated, horizontal stems that grow along the surface above ground. Other plants have *rhizomes*, which are thick, horizontal stems that grow above or below ground and send up *plantlets* at intervals. Many plants send up *plantlets* from their *roots*. Some plants produce *plantlets* on the surface, often along the margins, of their *leaves*. Under certain conditions, some plants produce *plantlets* on *flowering stems* where flowers would normally be formed.

Color the parent plants (A), fragments (J), and new plants (B) in the diagrams labeled "fragmentation."

Fragmentation is one of the commonest means of asexual reproduction by multicellular units. Fragmentation involves the physical separation, through various means, of a *parent plant* into a number of divisions or *fragments*. Each *fragment* is capable of forming a *new plant*.

One form of naturally occurring fragmentation is due to *senescence*, the natural aging and death of the older portions of a *parent plant*, of plants with surface stems that creep along a surface and root as they grow. Death of the older portions eventually divides the *parent plant* into a number of *fragments*, which continue to grow as independent plants.

Articulation, in which portions of a plant are adapted to break away from a *parent plant*, is another form of fragmentation. In some plants, an area of weakness is present between recent growth and older growth. Until maturation, which may take a year or more, strengthens these areas, recent growth may fall away from the *parent plant* when it is jarred. These articulated pieces, *fragments*, fall to the ground, where they may eventually root and form a *new plant*.

A common form of fragmentation is initiated by damage to a *parent plant*. In this type of fragmentation, a *parent plant* is broken into two or more *fragments* by some external force, such as water currents, wind, or animal activities. Each of the *fragments* produced may grow and produce a *new plant*.

MULTICELLULAR UNITS. PARENTₐ NEW PLANTв

GEMMАc

BULBILᴅ

ONION   NUTGRASS   LILY

LIVERWORT

PLANTLETв'

STRAWBERRY

MATERNITY PLANT

STOLONᴇ

MILKWEED

SPIDER
PLANT

BEACHGRASS   RHIZOMEᶠ   ROOTSɢ   LEAVESн   FLOWERING
STEM ɪ

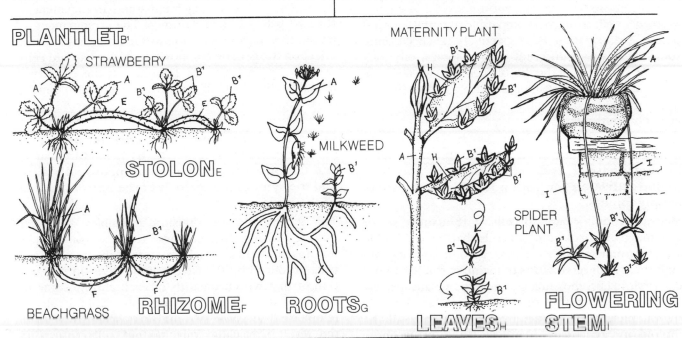

FRAGMENTⱼ
SENESENCE ★

GROUND
PINE

ARTICULATION ★

CHOLLA

DAMAGE ★

SPIROGYRA

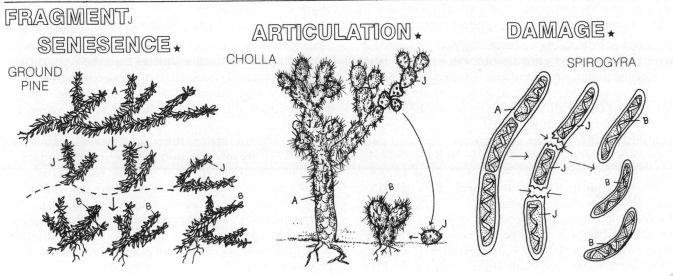

14
MEIOSIS

In the diploid cell at the top of the plate, color chromosome set A (A) and chromosome set B (B). Use two contrasting colors.

This plate illustrates the basic behavior of chromosome sets and the change in the chromosome complement, or chromosome set number, that occurs during meiosis (sexual cell division), a major event in the sexual reproduction cycle. The diploid cell at the top has two chromosome sets, *chromosome set A* and *chromosome set B,* from two compatible parental sources, parent A and parent B, respectively. Each chromosome set has two different chromosomes, one long and one short. Since the diploid cell has two chromosome sets, one from each compatible parent, it contains two pairs, one long and one short, of homologous chromosomes.

Color chromosome set A and chromosome set B in metaphase I and telophase I of the first meiotic division for both possible chromosome arrangements.

Meiosis produces haploid cells, with a single set of chromosomes, from diploid cells that have a double set of chromosomes. In sexual reproduction, diploid cells must undergo meiosis to maintain stable haploid and diploid chromosome numbers for a plant species. If the next generation of sex cells produced by the diploid phase were formed by mitosis, they would be diploid and genetically identical. Upon fusion, these diploid sex cells would form a tetraploid cell with four of each kind of chromosome (not shown). If this pattern continued, each succeeding generation would have twice as many chromosomes in each cell as the preceding generation. One major function of meiosis is to counterbalance the chromosome number doubling effect of sex cell fusion by reducing the chromosome level of the diploid phase by half. This is accomplished through the separation of the two chromosomes of each homologous pair in a diploid cell.

Meiosis consists of two sequential divisions. As in mitosis, duplicate strands of genetic material, DNA, are copied for each chromosome prior to division. Each chromosome then consists of two chromatids connected by a common centromere. The first meiotic division, or reductional division, separates the chromosomes of each homologous pair to form two

haploid cells. That is, one chromosome from each homologous pair goes to each haploid cell formed.

During metaphase I, the members of each homologous pair become arranged on either side of the equatorial plate, an imaginary plane through the middle of the cell (not shown), in a random manner so all the chromosomes of one parental chromosome set are not necessarily on the same side. Note the two possible arrangements, arrangement 1 and arrangement 2, for two pairs of chromosomes. The arrangement of paired chromosomes during metaphase I has a direct effect on the genetic composition of the haploid cells produced by meiosis. By telophase I, there are two haploid cells in which the chromosomes still consist of two duplicate chromatids.

Color the chromosomes from chromosome set A (A) and chromosome set B (B) in metaphase II and telophase II of the second meiotic division and the symbol for meiosis at the bottom center of the plate.

The second meiotic division resembles a typical mitotic division. In metaphase II, the chromosomes line up as individuals in the centers of the two haploid cells produced by the first meiotic division. Separation of the two chromatids of each chromosome in the two haploid cells produces four haploid cells in telophase II. Because two divisions occur in meiosis, the result is usually four haploid cells from one diploid parent cell. The genetic composition of the haploid cells on the left is identical to the parental sex cells, but the haploid cells produced by the chromosome arrangement at metaphase I on the right differ from the parental types. Each of these contains a mixture of parental chromosomes and has a genetic composition that differs from the individual parental sex cells.

As a second major function, meiosis provides a means for the introduction of genetic variation into a population. One form of genetic recombination, due to the random assortment of parental chromosomes in metaphase I, is examined on this plate in arrangements 1 and 2. Unlike mitosis, in which cells genetically identical to the parent cell are produced, meiosis produces cells that may exhibit a wide range of variation.

MEIOTIC CHROMOSOMAL BEHAVIOR.

CHROMOSOME SET A_A
CHROMOSOME SET B_B

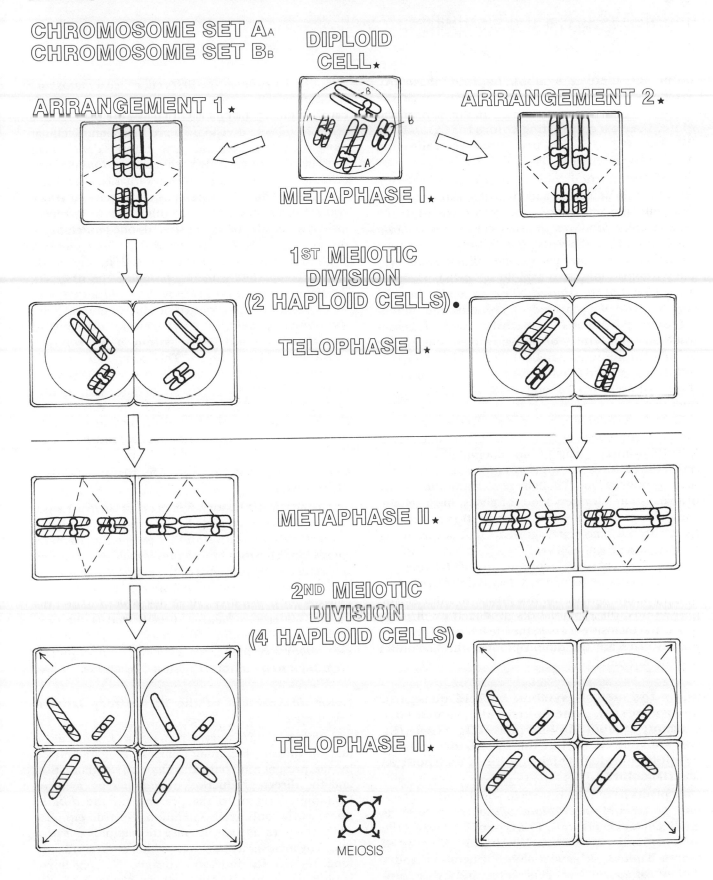

CHROMOSOME SET A₍ₐ₎
CHROMOSOME SET B₍ᵦ₎

DIPLOID CELL★

ARRANGEMENT 1★

ARRANGEMENT 2★

METAPHASE I★

1ST MEIOTIC DIVISION (2 HAPLOID CELLS)●

TELOPHASE I★

METAPHASE II★

2ND MEIOTIC DIVISION (4 HAPLOID CELLS)●

TELOPHASE II★

MEIOSIS

15
TYPES OF LIFE HISTORIES

Color the meiosis symbol, haploid phase (A), syngamy symbol, and diploid phase (D) of the small diagram labeled "sex" at the top of the plate. Use two contrasting colors for (A) and (D). (B) and (C) will be presented in following diagrams.

Sexual reproduction consists of two genetically important cellular events and two life history phases. One cellular event, meiosis (the meiotic division of a diploid cell), produces the first cell of the *haploid phase*, in which the cells contain only one of each kind of chromosome. The second cellular event, syngamy (the fusion of two haploid sex cells), produces the first cell of the *diploid phase*, in which the cells contain two of each kind of chromosome (homologous pairs). Meiosis counterbalances the chromosome-doubling effect of syngamy by reducing the diploid chromosome number by one-half. These two cellular events and two life history phases are the four major components of any kind of sexual reproduction.

Sexual reproduction is important to the survival of populations, especially in a changing environment, as it is the primary means of introducing genetic variability, the basic resource for evolution and adaptation, into a population. Though gene mutation is the primary source for new kinds of genes, most genetic variability is due to genetic recombination that results from chromosome rearrangements that occur during meiosis and syngamy.

Three types of sexual reproduction life histories, or life cycles, can be recognized, based on the presence of the *haploid phase, diploid phase*, or both as true generations; the type of cells produced by meiosis; and when meiosis occurs in the life history. The three life histories are the diplohaplontic, the haplontic, and the diplontic.

Color the meiosis symbol, haploid phase (A), meiospores (B), gametophytes (A¹), gametes (C), syngamy symbol, diploid phase (D), zygote (E), and sporophyte (D¹) of the life history labeled "diplohaplontic," which is colored with both (A) and (D) colors.

Most plants have a life history in which both the *haploid phase* and the *diploid phase* are present as multicellular generations. This type of life history, the diplohaplontic, is characterized by the presence of both a haploid, or gametophyte, generation, and a diploid, or sporophtye, generation, and by the production of *meiospores* by meiosis. The size and duration of the two generations may differ significantly, but both are typically multicellular. Meiosis of a diploid cell produces the first cell, called a *meiospore* (four are usually produced from each diploid mother cell), of the *haploid phase*. A *meiospore* undergoes a series of mitotic divisions to produce a multicellular *gametophyte* generation. By mitotic division of specialized haploid cells within a *gametophyte,* one or more sex cells, called *gametes* (that is, sperm and egg), are produced. Depending on species, a single *gametophyte* may produce both kinds of *gametes*, male and female, or, as illustrated, one *gametophyte* produces only male *gametes* and another *gametophyte* produces only female *gametes*. Syngamy, fusion of two compatible gametes, produces the first cell, the *zygote*, of the *diploid phase*. A *zygote* undergoes a series of mitotic divisions to produce a multicellular *sporophyte* generation. By meiotic division of specialized diploid cells within a *sporophyte, meiospores* are produced. This completes one full cycle of the diplohaplontic life history. "Diplohaplontic" refers to the presence of both the *haploid phase* and the *diploid phase* as true generations. Because of this, the phrase "an alternation of generations" is also used to describe this life history.

Color all features of the life history labeled "haplontic."

The haplontic life history, which occurs in many algae and fungi, is characterized by the presence of only the *gametophyte* generation and the production of *meiospores* directly from the *zygote* by meiosis. "Haplontic" refers to the presence of the *haploid phase* as the only true generation. No true *sporophyte* generation exists because the first cell of the *diploid phase*, the *zygote,* undergoes meiosis directly and is the only diploid cell. Depending on species, one multicellular *gametophyte* or numerous unicellular *gametophytes* are produced by mitotic division of each *meiospore*.

Color all features of the life history labeled "diplontic."

The diplontic life history, which occurs in a few algae (and all animals, including humans), is characterized by the presence of only the *sporophyte* generation and the direct production of *gametes* by meiosis. "Diplontic" refers to the presence of the *diploid phase* as the only true generation. No true *gametophyte* generation exists because the haploid cells produced by meiosis are *gametes*, which are around only long enough to undergo syngamy. No vegetative haploid stage exists. Multicellular *sporophytes* are typically formed.

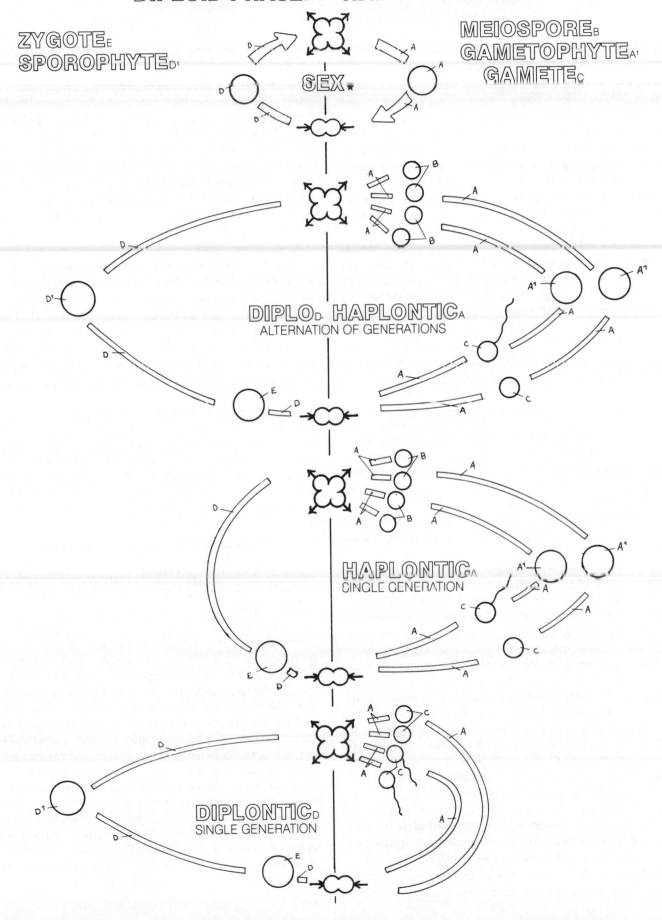

COMPONENTS OF LIFE HISTORIES

Color the isogamous (A), anisogamous (B), and oogamous (C) gamete types at the top of the plate.

Differences in the timing of events, in structure, and in function provide for numerous variations on the general formats of the three basic life cycles. One source of life cycle variation lies in the types of gametes, or sex cells, produced. Gametes are classified according to motility, size, and structural similarity. A plant species is either *isogamous, anisogamous,* or *oogamous,* depending on the types of gametes it produces.

In isogamous plants, the gametes, called isogametes, of both mating strains are identical in motility, size, and shape even though they are of two different mating strains. *Isogamous* (meaning "like gametes") is in reference to the identical appearance of the isogametes, which are usually designated as plus (+) and minus (−) strains rather than male and female. Most isogametes are motile by one or more flagella, but nonmotile isogametes (not shown) are also found. In most isogamous plants, syngamy occurs in open water away from the parental source.

Anisogamous plants produce gametes, called anisogametes, that are similar in appearance but not identical. Most differ slightly in size. *Anisogamous* (meaning "unlike gametes") is in reference to the dissimilar appearance of the two sexes of gametes. The larger gamete type is considered to be the female and the smaller, the male. Both anisogametes are usually motile by one or more flagella, but in some anisogamous plants, both anisogametes are nonmotile (not shown). As with isogametes, syngamy usually occurs in open water away from the parental source.

In oogamous plants, the gametes differ significantly in size, in structure, and often in motility. *Oogamous* (meaning "egg gametes") is in reference to the large female gamete, usually called an egg, which is nonmotile and much larger than the smaller gamete, usually called a sperm. Sperm are motile, by one or more flagella, in most oogamous plants, but some plants produce nonmotile sperm. In most oogamous plants, the egg remains in place on the parental plant and syngamy occurs on the parental plant, but some release the egg into open water. Whether motile or nonmotile, the sperm is typically released into open water except in the more specialized land plants.

Color the components of the homothallic (D) and heterothallic (E) source of gametes, including the gametes. Anisogametes (B) are illustrated.

Based upon the source of both gamete types, a plant species is classified as either *homothallic* or *hetero-*

thallic. If both gamete types, plus (+) and minus (−) or male and female, are produced by a single parental plant, the plant species is *homothallic*. Homothallic (meaning "same body") is in reference to the production of both gamete types by a single plant. If the two different gamete types are produced by two different parental plants, each of which produces only one gamete type, the plant species is *heterothallic*. Heterothallic (meaning "different body") is in reference to the production of each gamete type by a different plant.

Color the components of the isomorphic (F) and heteromorphic (G) alternation of generations.

In the diplohaplontic life cycle, in which a true alternation of generations occurs, two different alternations of generations are found. If the sporophyte and gametophyte generations are morphologically similar, as in *Ulva,* so as to be relatively indistinguishable in appearance, an *isomorphic* alternation of generations exists. *Isomorphic* (meaning "same form") is in reference to the identical superficial appearance of the two generations. If the sporophtye and gametophyte generations differ significantly in appearance, as in *Laminaria,* so as to be readily distinguishable, a *heteromorphic* alternation of generations exists. *Heteromorphic* (meaning "different form") is in reference to the dissimilar appearance of the two generations.

Color the gametophytes, bearing either male (H) or female (I) gametangia or both, produced by homosporous (J) and heterosporous (K and L) plants, which produce either microspores (K) or megaspores (L).

A particular plant species is either *homosporous* or *heterosporous,* depending on the type of meiospores it produces. *Homosporous* plants produce meiospores that are morphologically identical. In some plants, the gametophytes produced from homosporous meiospores are bisexual and produce both male and female gametangia; in other plants, they are unisexual and produce either male or female gametangia, but not both. In *heterosporous* plants, the meiospores differ in size and sex. The small meiospores, called *microspores,* produce male gametophytes that form only male gametangia; the large meiospores, called *megaspores,* produce female gametophytes that form only female gametangia. Homosporous plants are either *homothallic* or *heterothallic,* but all *heterosporous* plants are *heterothallic.* Most land plants are *heterosporous.*

LIFE HISTORY VARIATIONS.

GAMETE TYPES ★

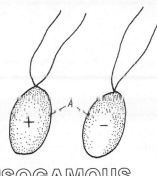 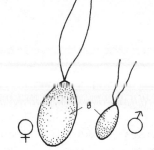

ISOGAMOUS_A   ANISOGAMOUS_B   OOGAMOUS_C

GAMETE SOURCE ★

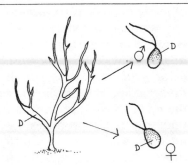 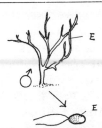 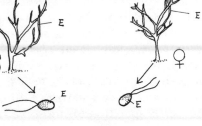 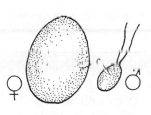

HOMOTHALLIC_D   HETEROTHALLIC_E

ALTERNATION OF GENERATIONS ★

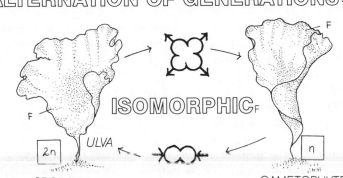

ISOMORPHIC_F

ULVA

2n — SPOROPHYTE   n — GAMETOPHYTE

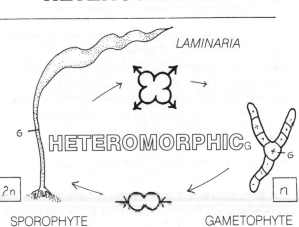

LAMINARIA

HETEROMORPHIC_G

?n — SPOROPHYTE   n — GAMETOPHYTE

HOMOSPORY AND HETEROSPORY ★

GAMETOPHYTE ★
  MALE_H
  FEMALE_I
HOMOSPOROUS_J
HETERO_K SPOROUS_L
  MICROSPORE_K
  MEGASPORE_L

 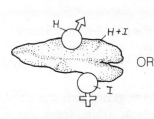 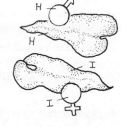

OR

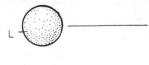 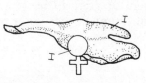

17
PLANT GENETICS

Color the nucleus (A), chromosomes (B), centromere (C), gene loci (D — use gray), diploid nucleus (E), and haploid nucleus (F). Stop with the haploid nucleus.

The *nucleus* contains the *chromosomes* that during asexual or sexual reproduction function as the carriers of heritable traits, or characteristics, from generation to generation. Each *chromosome* can be abstractly pictured as a string of beads with each bead position equivalent to a specific *gene locus*. The position of each *gene locus*, and that of the *centromere*, is constant for a specific *chromosome*. At each *gene locus* is a specific gene that contains the genetic code for a particular trait. A trait may be functional, such as a digestive enzyme, or it may be physical, such as flower color.

The two homologous *chromosomes* of each chromosome pair in a *diploid nucleus* have a matching sequence of *gene loci*. Therefore, the flower color *gene locus* is at the same position on each member of the homologous *chromosomes* with this *gene locus*. Two specific flower color genes are present, one at the flower color *gene locus* of each homologous *chromosome*. A *haploid nucleus* contains only one of each kind of *chromosome* and therefore a single flower color *gene locus* and flower color gene. Hundreds of different *gene loci* are present on each *chromosome*, but for clarity, a limited number are depicted in the hypothetical examples illustrated. Genotype refers to the genetic composition, or specific genes present, and phenotype refers to the physical expression of the genes.

Color the red alleles (G) red, but leave the white alleles (H) uncolored in the two diploid chromosome sets labeled "homozygous." Also color one flower (G¹).

A *gene locus* may have genes that code for different expressions of the same trait. These are known as alleles. Usually, only two *alleles* for a specific *gene locus*, one on each member of a homologous chromosome pair, are present in an individual *diploid* plant.

In a hypothetical plant, flower color is determined by the *alleles* present at *gene locus* 4, next to the *centromere*. Two different flower color *alleles*, red

and *white*, are present within the gene pool of the hypothetical plant population. The *red allele* contains the genetic information for the production of *red flowers* and the *white allele*, that for the production of *white flowers*. If the *alleles* at a specific *gene locus* are identical on homologous *chromosomes*, the plant is said to be *homozygous* for the trait encoded by the *alleles*. Thus, if the *red allele* is present at both flower color *gene loci*, the plant is *homozygous* for *red flower* color and produces *red flowers*. Similarly, if the *white allele* is present in the *homozygous* condition, the plant produces *white flowers*. The two *alleles*, *red* and *white*, are different forms of the same flower color gene, and when present in the *homozygous* condition, they produce different physical expressions, or phenotypes.

Color the red allele (G) and the flowers (G¹ and I) for the diagrams labeled "heterozygous."

When two different *alleles* are present at a specific *gene locus* on two homologous *chromosomes*, the plant is said to be *heterozygous* for the trait expressed. Thus, with a *red allele* on one chromosome and a *white allele* on the other, the plant is *heterozygous* for flower color.

When different *alleles* are present, one *allele* may be *dominant* over the other and so have total control over genetic expression. Therefore, if both a *red allele* and a *white allele* are present and the *red allele* is *dominant*, the plant will produce *red flowers* even though the *white allele* is present in the genotype. The *white allele* is said to be *recessive* because it has no influence in determining flower color. Traits coded by *recessive alleles* are expressed only in the *homozygous* condition.

Some *alleles* do not express *dominance* or *recessiveness*. Each of the two *alleles* present has some input into genetic expression. In the hypothetical plant, the flower color produced would be a blend. A *red allele* and a *white allele* on the homologous chromosome pair would yield *pink flowers*. Input by both *alleles* present at a specific *gene locus* on homologous *chromosomes* is called *codominance*. It is also known as incomplete dominance and blending inheritance.

INTRODUCTION TO PLANT GENETICS.

NUCLEUS_A — rendered as NUCLEUS$_A$

NUCLEUS$_A$
CHROMOSOME$_B$
CENTROMERE$_C$
GENE LOCUS$_D$ ★
DIPLOID NUCLEUS$_E$
HAPLOID NUCLEUS$_F$
ALLELES ★
    RED$_G$
    WHITE$_H$ -¦-
FLOWER COLOR ★
    RED$_{G^1}$
    WHITE$_{H^1}$ -¦-
    PINK$_I$

DIPLOID$_E$

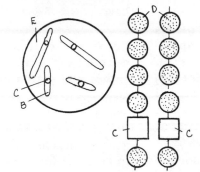

HAPLOID$_F$

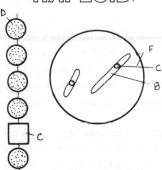

HOMOZYGOUS ★

HETEROZYGOUS ★

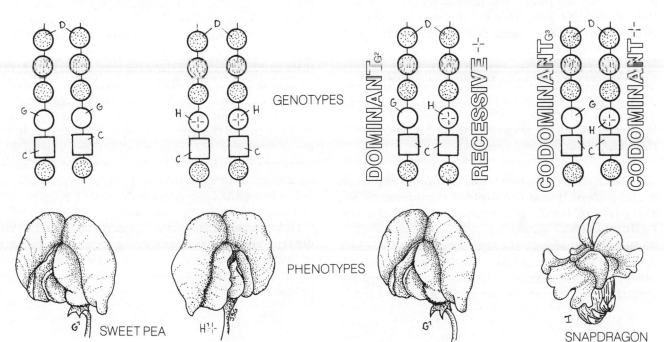

GENOTYPES

DOMINANT$_{G^2}$ RECESSIVE -¦-   CODOMINANT$_{G^3}$ CODOMINANT -¦-

PHENOTYPES

G^1 SWEET PEA H^1 -¦- G^1 I SNAPDRAGON

A monohybrid cross is that between parents homozygous for two different *alleles* at a single *gene locus*. This plate combines two monohybrid crosses in order to include the effects of gene linkage, which is the constant association of all *gene loci* of one chromosome. The two linked traits examined are flower color and plant height in a hypothetical plant in which the *alleles* for flower color exhibit codominance and the *alleles* for height exhibit dominance. *Centromeres* are included as a reference point.

Color the homologous chromosomes and plants for parent A and parent B. Use gray for unspecified gene loci (A) and black for centromeres (B). Use red for red alleles (C) and red flowers (C¹) and a shade of green for tall alleles (D) and tall stems (D¹). Leave white alleles (E) and white flowers (E¹) uncolored, and use a second shade of green for dwarf alleles (F) and dwarf stems (F¹).

Diploid *parent A* is homozygous for flower color and height. It has two *red alleles* (R) that code for *red flowers* and two *tall alleles* (T) that code for *tall stems*. Therefore, the genotype of *parent A* is RRTT because only *red alleles* and *tall alleles* are present at their respective *gene loci* on the two homologous chromosomes. This genetic composition gives parent A a *red-flowered, tall* phenotype.

Diploid *parent B* is also homozygous at *loci* for both flower color and height. It has two *white alleles* (r) that code for *white flowers* and two *dwarf alleles* (t) that code for *dwarf stems*. Because *parent B* is homozygous, its genotype is rrtt, and it has a *white-flowered, dwarf* phenotype.

Color the chromosomes of the gametes of parent A and parent B and the diploid chromosome set and plant of the F₁ generation. Use pink for the F₁ flowers (G).

Because both parents are homozygous, the gametes produced by *parent A* have only *red alleles* and *tall alleles* and an RT genotype because they are haploid. Similarly, the gametes of *parent B* have only *white*

alleles and *dwarf alleles* and an rt genotype. Fusion of gametes from *parent A* and *parent B* forms the diploid F₁, or first, generation. All members of the F₁ generation are heterozygous for both *gene loci* and have an RrTt genotype because both *red* and *white* flower color *alleles* and *tall* and *dwarf* stem *alleles* are present in their genotype. Given that the *red allele* exhibits codominance with the *white allele* and the *tall allele* is dominant over the *dwarf allele* in the hypothetical plant, all members of the F₁ generation have a *pink-flowered, tall* phenotype.

Color the chromosomes of the gametes produced by the F₁ generation and the chromosomes and plants of the F₂ generation.

Members of the F₁ generation, F₁ parents, are crossed, gametes are exchanged, to produce the F₂, or second, generation of a monohybrid cross. Because the F₁ generation is heterozygous and the genes for the two traits are on the same chromosome, only two different gamete genotypes, RT and rt, are produced. Either of the two gamete types for eggs can be fertilized by either of the two gamete types for sperm; so four different combinations of homologous chromosomes, and genotypes, are possible in the F₂ generation.

Two of the possible combinations produce heterozygous progeny. The remaining possible combinations produce two homozygous types. Therefore, the ratio of genotypes is 1 RRTT : 2 RrTt : 1 rrtt. Match the different gamete types to confirm this ratio, which is typical for a monohybrid cross. The phenotypic ratio may not match the genotypic ratio. Where the alleles are codominant, as in flower color, the genotypic and phenotypic ratios are the same. But where an allele is dominant, as in stem height, the phenotypic ratio is 3:1 because the homozygous dominant progeny and heterozygous progeny look alike.

Linked genes are inherited together because they are not separated from one another during meiosis. Because of linkage, white-flowered, tall plants; pink-flowered, dwarf plants; and red-flowered, dwarf plants are not produced in the hypothetical monohybrid cross.

MONOHYBRID CROSS WITH LINKAGE.

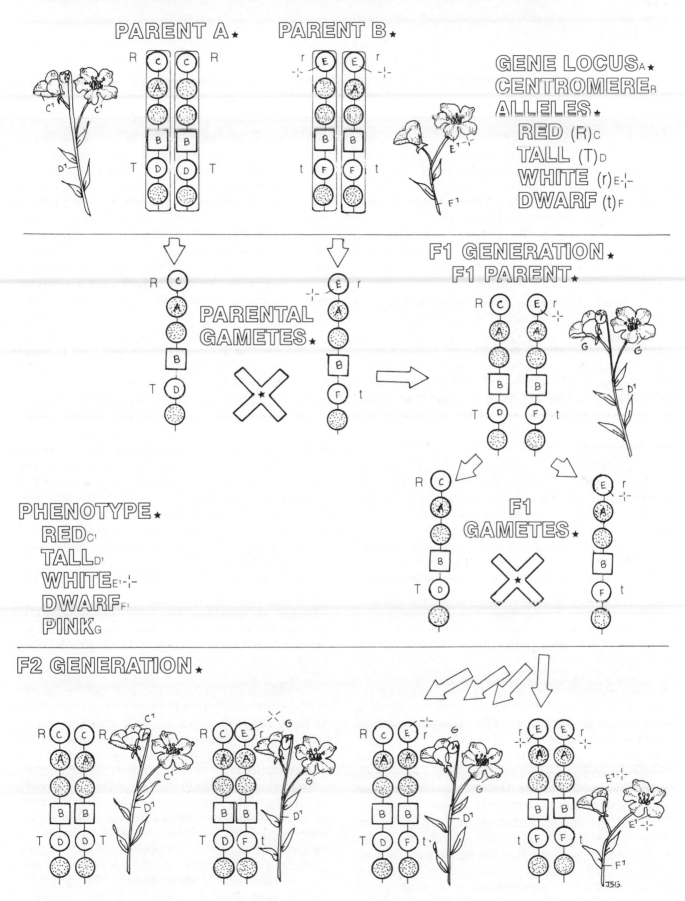

PARENT A. PARENT B.

GENE LOCUS A.
CENTROMERE B
ALLELES.
RED (R) C
TALL (T) D
WHITE (r) E
DWARF (t) F

PARENTAL GAMETES.

F1 GENERATION.
F1 PARENT.

F1 GAMETES.

PHENOTYPE.
RED C¹
TALL D¹
WHITE E¹
DWARF F¹
PINK G

F2 GENERATION.

JSG.

PLANT GENETICS

A dihybrid cross is a cross between parents homozygous for two different alleles at two *gene loci* on two different homologous chromosome pairs. It demonstrates how alleles at *gene loci* on separate chromosomes are inherited. The two independently segregating traits examined on this plate are seed coat texture and seed color in peas. The alleles that code for both of these traits exhibit dominance in peas. *Centromeres* are included as a reference point.

Color the homologous chromosomes and peas for parent A and parent B. Use gray for unspecified gene loci (A) and black for centromeres (B). Color the smooth alleles (C) and smooth seed coat (C¹), the yellow alleles (D) and yellow seeds (D¹), the wrinkled alleles (E) and wrinkled seed coat (E¹), and the green alleles (F) and green seeds (F¹). Use yellow and green for the appropriate seed colors. Choose any colors for seed coat textures.

Diploid parent A is homozygous at its *gene loci* for seed coat texture and seed color. It has two dominant *smooth alleles* (S) that code for *smooth seed coat* and two dominant *yellow alleles* (Y) that code for *yellow seed color* in pea seeds (peas). Therefore, the genotype of parent A is SSYY because only *smooth alleles* and *yellow alleles* are present at their respective *gene loci* on the two different homologous chromosome pairs. This genetic composition gives parent A a *smooth, yellow-seeded* phenotype.

Diploid parent B is also homozygous at both seed coat texture and seed color *loci* on the two different homologous chromosome pairs. It has two recessive *wrinkled alleles* (s) that code for *wrinkled seed coat* and two recessive *green alleles* (y) that code for *green seed color*. Because parent B is homozygous recessive, its genotype is ssyy, and it has the recessive *wrinkled, green-seeded* phenotype.

Color the chromosomes of the gametes of parent A and parent B and the diploid chromosome set and pea of the F_1 generation.

Since both parents are homozygous, the gametes produced by parent A have only *smooth alleles* and *yellow alleles* and an SY genotype because they are haploid. Similarly, the gametes of parent B have only *wrinkled alleles* and *green alleles* and an sy genotype. Fusion of gametes from parent A and parent B forms the diploid F_1, or first, generation. All members of the F_1 generation are heterozygous at both *gene loci* and have an SsYy genotype because *smooth* and *wrinkled* seed coat *alleles,* as well as *yellow* and *green* seed color *alleles,* are present in their chromosome set. Because *smooth alleles* and *yellow alleles* are dominant, all peas produced by the F_1 generation have a *smooth, yellow-seeded* phenotype. No genotypic or phenotypic variation is present in the F_1 generation, but the heterozygous condition of two *gene loci* on separate chromosomes is a basis for chromosome rearrangements during metaphase I in gamete formation.

Color the chromosomes of the gametes produced by the F_1 generation and the chromosomes and peas of the F_2 generation found within the Punnett Square.

Members of the F_1 generation are crossed to form the F_2, or second, generation of a dihybrid cross. Because seed coat texture and seed color alleles are on different homologous chromosome pairs, they assort independently of one another during metaphase I of meiosis, and F_1 gametes of four different genotypes (SY, Sy, sY, and sy) are produced. Since any of the four possible gamete types for eggs can be fertilized by any of the four possible gamete types for sperm, sixteen different combinations of homologous chromosomes, or genotypes, are possible in the F_2 generation. Some of these, as in the monohybrid cross, are duplicate combinations. The different genotypes can be determined by use of a Punnett Square, in which the gamete types for eggs are arranged down one side and the gamete types for sperm are arranged along the top. To obtain the genotypes, sum the two gamete types within each box where they intersect. Naturally occurring genotypic and phenotypic ratios should approximate the ratios obtained using the Punnett Square since there is an equal probability that any two gamete types will fuse. Note that two new phenotypes appear in the F_2 generation because of recombination of parental chromosomes.

This example illustrates the introduction of genetic variability into a population through rearrangements of chromosomes during meiosis and syngamy. Two *gene loci* were examined in this example, but hundreds of *gene loci* are present on each chromosome, and diploid chromosome numbers range from four to hundreds in plant species. This type of recombination alone accounts for a large amount of genetic variation.

DIHYBRID CROSS.

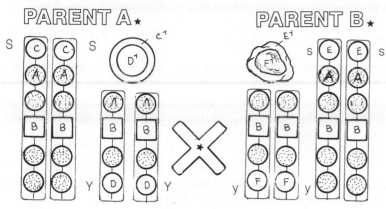

PARENT A ★ PARENT B ★

GENE LOCUS A ★
CENTROMERE B
ALLELES ★
   SMOOTH (S) C
   YELLOW (Y) D
   WRINKLED (S) E
   GREEN (y) F

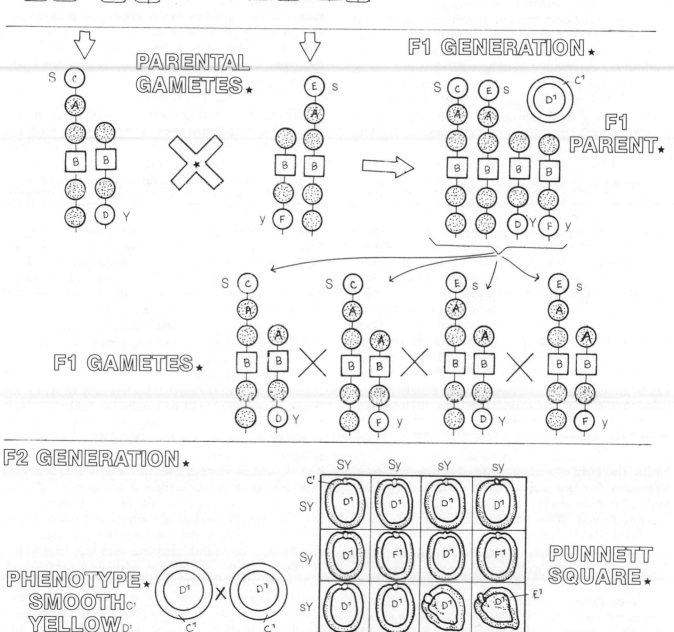

PARENTAL
GAMETES ★

F1 GENERATION ★

F1
PARENT ★

F1 GAMETES ★

F2 GENERATION ★

SY   Sy   sY   sy

PHENOTYPE ★
SMOOTH C¹
YELLOW D¹
WRINKLED E¹
GREEN F¹

PUNNETT
SQUARE ★

20
PLANT GENETICS

Color chromosome A (A) and chromosome B (B) in the diploid cell formed by fusion and in the row of sticklike chromosomes immediately below it. Do not use red for either chromosome. Color the centromere (C) black.

Crossing-over, a major means of introducing genetic variation into a population, involves the exchange of genetic material between chromatids of homologous chromosomes during prophase I of meiosis. In the hypothetical example illustrated, fusion of one haploid sex cell from parent A, with *chromosome A,* with one from parent B, with *chromosome B,* forms a diploid cell with a single pair of homologous *chromosomes, A* and *B.* During prophase I, homologous chromosomes, each consisting of two chromatids held together by a common *centromere,* pair with one another. While paired, a chromatid from one member of a homologous pair often overlaps (crosses over) an equal portion of a chromatid from the other member of the homologous pair. This is frequently accompanied by separation of the chromatid portions distal to the point of crossover from their parent chromosomes and the reattachment of these separated chromatid pieces to the broken end on the opposite chromosome parent. In the example, a portion of a chromatid from parent A's chromosome has become reattached to parent B's chromosome, and an equal portion of chromatid from parent B's chromosome has become attached to parent A's chromosome. The result of this exchange of chromosome portions is an exchange of genetic material and the possible introduction of genetic variation. How does this introduce variation? After all, *chromosome A* and *chromosome B* are homologous.

Color the row of string-of-beads chromosomes. Use gray for the unspecified gene loci (D) and red for red alleles (E), and leave the white alleles (F) uncolored. Also, color the smooth (G) and serrated (H) alleles.

Consider the effects of crossing-over in a hypothetical plant in which two specific gene loci that code for two different traits, flower color and flower edge texture, are examined. In the example, the *red allele* (R) is dominant over the *white allele* (r), and the *smooth allele* (S) is dominant over the *serrated allele* (s). The plant is heterozygous for both traits and therefore has an RrSs genotype and a red, smooth-flowered phenotype. *Chromosome A* has a *red allele* at the flower color gene locus and a *smooth allele* at the flower edge texture gene locus, and *chromosome B* has a *white allele* and a *serrated allele.* Because the *red* and *smooth alleles* are linked (located on the same chromosome), they would be expected always to be inherited together. Similarly, the *white* and *serrated alleles* should be inherited together due to linkage. In nature, crossing-over often disrupts linkage. Observe the difference in genetic composition between the two chromatids of each homologous chromosome following crossing-over in the string-of-beads chromosome models.

Color the different gamete types illustrated and the flower phenotypes within the Punnett Square. Use red for red flowers (E^1), and leave the white flowers (F^1) uncolored. Use the same colors for the smooth edges (G^1) and serrated edges (H^1) as you used for the smooth (G) and serrated (H) alleles.

With no crossing-over, the only gamete types the hypothetical plant would produce would be RS and rs, and the expected genotypic ratio produced by a cross of these gametes would be 1 RRSS : 2 RrSs : 1 rrss. The phenotypic ratio would be 3 *red, smooth-flowered* plants : 1 *white, serrated-flowered* plant. However, crossing-over alters the expected ratios for linked gene loci by rearranging genetic material between homologous chromosomes. Following crossing-over, meiosis results in two gamete types, Rs and rS, in addition to the two expected gamete types. All four gamete types will be present in both male and female gametes. Since crossing-over occurs in a relatively small percentage of meiotic divisions, only a small percentage of crossover gamete types (that is, Sr and sR) are produced. Due to the low frequency of crossover gamete types, there is not an equal chance of fusion between any two of the four gamete types present. Therefore, the genotypes and phenotypes, but not their ratios, may be determined using a Punnett Square. From this, crossing-over may lead to the formation of new genotypic and phenotypic combinations, or variation, in following generations. Without crossing-over, only *red, smooth-flowered* and *white, serrated-flowered* plants would be present in the population. But *red, serrated-flowered* and *white, smooth-flowered* plants are possible with crossing-over.

CHROMOSOME CROSSING-OVER.

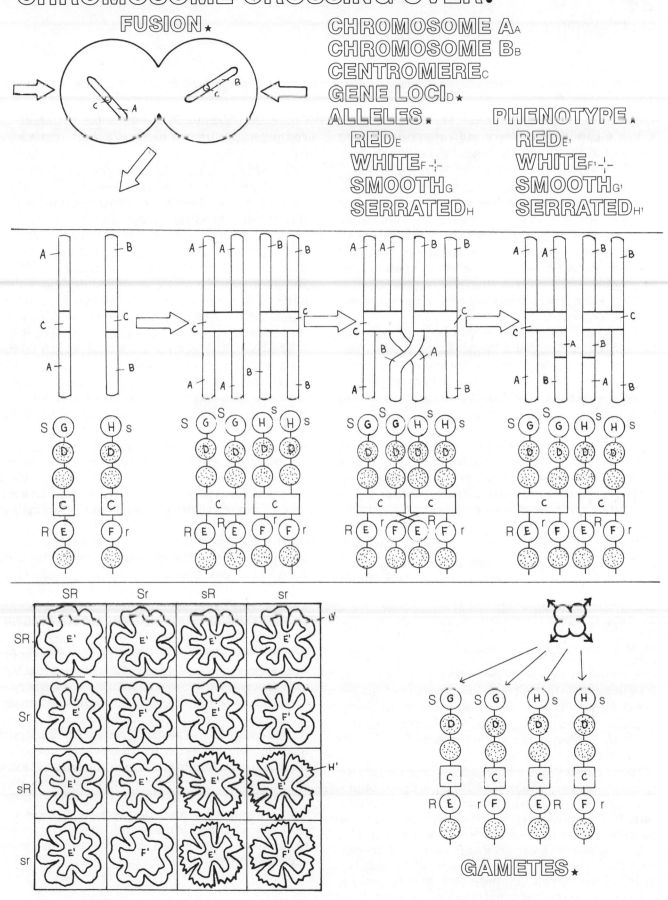

FUSION ★

CHROMOSOME A_A
CHROMOSOME B_B
CENTROMERE_C
GENE LOCI_D ★

ALLELES ★	PHENOTYPE ★
RED_E	RED_{E'}
WHITE_F -¦-	WHITE_{F'} -¦-
SMOOTH_G	SMOOTH_{G'}
SERRATED_H	SERRATED_{H'}

GAMETES ★

LOWER PLANT GROUPS

Color the diagrams of bacteria (A) and blue-green algae (B). Numbers indicate approximate number of species in each group, and arrows indicate a hypothetical evolutionary sequence.

Bacteria, Schizomycophyta, are ubiquitous in the environment. Most are minute unicellular forms, but some are filaments and simple colonies. Both motile and nonmotile vegetative cells are found. Cell walls contain mucopeptides. A few are photosynthetic, and some are parasitic, but most are saprophytic. No major pigments are present. Bacterial reproduction is asexual, but nonsexual mechanisms for genetic materials transfer exist.

Blue-green algae, Cyanophyta, are primarily aquatic, especially in marine habitats, as either free-floating or surface-inhabiting forms, or terrestrial, especially in continually moist substrates rich in nitrates. Unicellular and filamentous forms and some simple colonies exist. No motile cells are found. Cell walls contain cellulose, pectins, and proteins, and most are surrounded by a gelatinous sheath. All are photosynthetic, and major pigments are chlorophyll a, carotenoids, phycocyanin, phycoerythrin, and xanthophylls. Reserve food materials are stored as glycogen and proteins. Reproduction is strictly asexual.

Color the diagrams of fungi (C) and lichens (D).

Fungi, Myxomycophyta (slime molds) and Eumycophyta (true fungi), are a large and relatively unknown group. Like *bacteria* and *blue-green algae*, they are found in many different habitat types. Most are multicellular filamentous forms, but there are some unicellular forms. Most have no motile vegetative cells, though slime molds are a distinct exception. Cell walls contain chitin or cellulose or both. None are photosynthetic, and most are saprophytic, though a few are infectious parasites on plants and animals. No major pigments are present. Both asexual and sexual reproduction occur in most members, and some produce motile sex cells.

Lichens are a single plant that is a composite of two species from two different divisions, a photosynthetic green or *blue-green algae* and a saprophytic *fungus*, that grow in close association. *Lichens* grow attached to surfaces, such as rocks and tree trunks, and from the seashore to mountaintops to deserts. Because of their composite structure, all are multicellular. No motile cells are produced. Reproduction is commonly asexual by fragmentation, but some specialized sexual reproduction occurs.

Color the diagrams of golden-brown algae (E), red algae (F), brown algae (G), and green algae (H).

Golden-brown algae, Chrysophyta, are abundant in both marine and freshwater habitats as either free-floating or attached forms. Most are unicellular, though some form simple colonies. Motile and nonmotile vegetative cells are found. Cell walls contain cellulose and pectin; those of diatoms contain silica. All are photosynthetic, and major pigments are chlorophylls a and c, carotenoids, and xanthophylls. Reserve food materials are stored as modified starch, fats, and oils. Reproduction is by asexual and sexual means, and some produce motile sex cells.

Red algae, Rhodophyta, are predominantly marine, but a few freshwater and terrestrial forms exist. They range from unicellular forms to sheetlike or highly branched filamentous forms of small size. No motile cells are produced. Cell walls contain cellulose, pectin, polysaccharides, and calcium carbonate. Most are photosynthetic, and major pigments are chlorophyll a, carotenoids, phycocyanin, phycoerythrin, and xanthophylls. Reserve food materials are stored as modified starch (floridean starch). Reproduction is by asexual and sexual means, and no motile sex cells are produced.

Brown algae, Phaeophyta, are almost exclusively marine in temperate climates. No unicellular forms are known, and most are filamentous, from microscopic size to huge kelps. Cell walls contain cellulose, pectin, alginic acid, and fucoidin. Most are photosynthetic, and major pigments are chlorophylls a and c, carotenoids, fucoxanthin, and xanthophylls. Reserve food materials are stored as laminarin starch. Both asexual and sexual reproduction occur. Only sex cells are motile.

Green algae, Chlorophyta, are widespread. Most are aquatic in fresh water, but some marine and terrestrial forms exist. They are widely diverse and range from unicellular forms to multicellular forms and some colonies. Motile and nonmotile vegetative cells are found. Cell walls contain mostly cellulose. All are photosynthetic; major pigments are chlorophylls a and b, carotenoids, and xanthophylls. Reserve food materials are stored as true starch. Reproduction is by asexual and sexual means, and most produce motile sex cells.

LOWER PLANT GROUPS.

PROKARYOTIC★
   BACTERIA_A
   BLUE-GREEN ALGAE_B
EUKARYOTIC★
   FUNGI_C
   LICHENS_D
   GOLDEN-BROWN ALGAE_E
   RED ALGAE_F
   BROWN ALGAE_G
   GREEN ALGAE_H

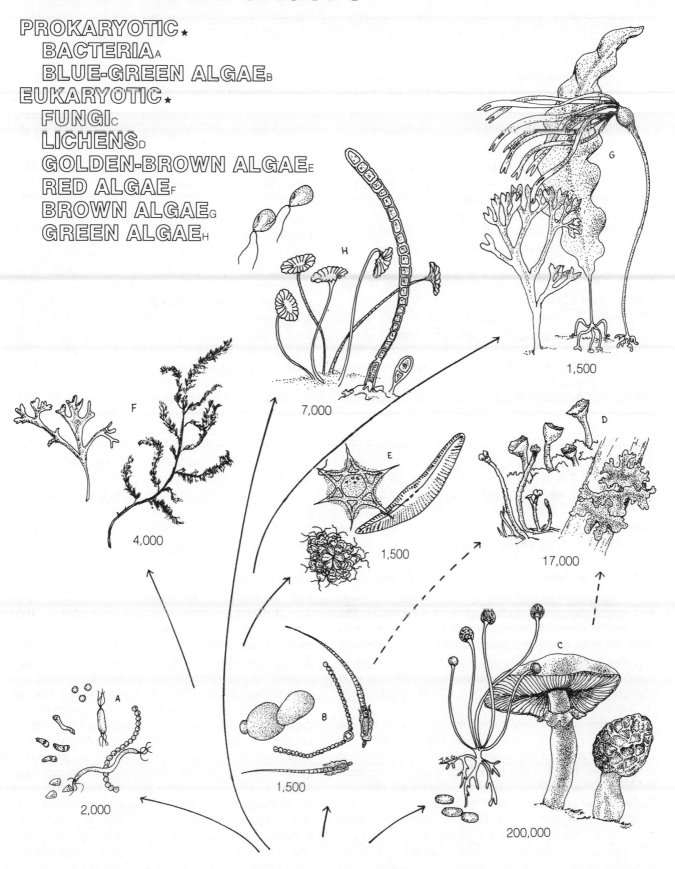

Increased structural and physiological complexity is required for land plants, primarily because the environmental conditions are harsher and less stable than those of the aquatic environment.

All land plants have diplohaplontic life histories and, therefore, a true alternation of generations.

Color the bryophyte representatives (A) in the lower right corner.

The mosses, liverworts, and hornworts (the *bryophytes,* division Bryophyta) are the simplest of the green land plants and are nonvascular. Though their distribution is worldwide, they occur most commonly in continually moist habitats of the temperate regions. All *bryophytes* are homosporous. The gametophyte, the most visible generation, is free-living; the sporophyte is typically partially or completely dependent (parasitic) upon the gametophyte. *Bryophytes* have few adaptations for support or for conduction of water, minerals, and food. As a result, *bryophytes* are usually rather small and tend to grow appressed to or not far above the substrate. Though some *bryophytes* occur in arid habitats, all require abundant water for gamete transfer.

Color the representative of the ferns and fern allies (B) just above the bryophytes (A).

The *ferns,* division Pterophyta, and *fern allies,* the psilopsids, division Psilophyta, the club mosses and spike mosses, division Lycophyta, and the horsetails, division Arthrophyta, constitute a large, widespread assemblage with close ties to many ancient plants well represented in the fossil record. These are the non-seed-bearing vascular plants. *Ferns and fern allies* range from aquatic to desert habitats, but they most commonly occur in areas of continually abundant moisture. The sporophyte generation of the *ferns and fern allies* is the most visible generation. Both homosporous and heterosporous species are present, but most are homosporous. The gametophyte generation is an independent, small, inconspicuous plant that has few adaptations for support and conduction. The sporophyte generation, which soon becomes independent after an initial period of dependence on the gametophyte, is often a large, conspicuous plant with adaptations for support and conduction that allow an upright growth habit. As with the *bryo-*phytes, the *ferns and fern allies* require abundant water for gamete transfer.

Color the gymnosperm representatives (C) on the left side of the plate.

The largest group of *gymnosperms* are the cone-bearing plants, or conifers, division Coniferophyta; but included in this assemblage are the cycads, division Cycadophyta, *Ginkgo,* division Ginkgophyta, and *Gnetum* and *Welwitschia,* division Gnetophyta. This is the least specialized assemblage of seed-bearing vascular plants. *Gymnosperm,* meaning ''naked seeds'' refers to the production of seeds that develop exposed to the atmosphere. *Gymnosperms* as a group are found from the temperate regions to the tropics. Habitats vary from rather moist environments to the arid environment of deserts. The sporophyte generation is prominent, with well-developed adaptations for support and conduction. All species are heterosporous. The female gametophyte is parasitic upon the sporophyte, and the male gametophyte is a minute plant surrounded by a thick spore wall. Conifers do not require water for gamete transfer, and most do not produce motile sperm.

Color the representatives of the flowering plants (D) in the top right corner.

The *flowering plants,* division Anthophyta, are the most successful, diverse, and widespread of the higher plant groups. The flower, a structure highly specialized for sexual reproduction, is a characteristic feature of members of this group. Flowering plants are cosmopolitan in all types of habitats. They have well-developed systems for support and conduction and may attain large size. Growth habits vary from small plants less than a centimeter in diameter to trees exceeding three hundred fifty feet in height. Like the *gymnosperms,* the *flowering plants* are all heterosporous. The male gametophyte is an encapsulated plant with stored food reserves similar to that of *gymnosperms.* The female gametophyte is parasitic upon the sporophyte and surrounded by sporophytic tissues. The developing seeds of *flowering plants* are completely surrounded by sporophytic tissue and are not exposed to the atmosphere. No motile sperm are produced, and water is not required for gamete transfer.

GREEN LAND PLANT GROUPS.

BRYOPHYTESA
FERNS AND FERN ALLIESB
GYMNOSPERMSC
FLOWERING PLANTSD

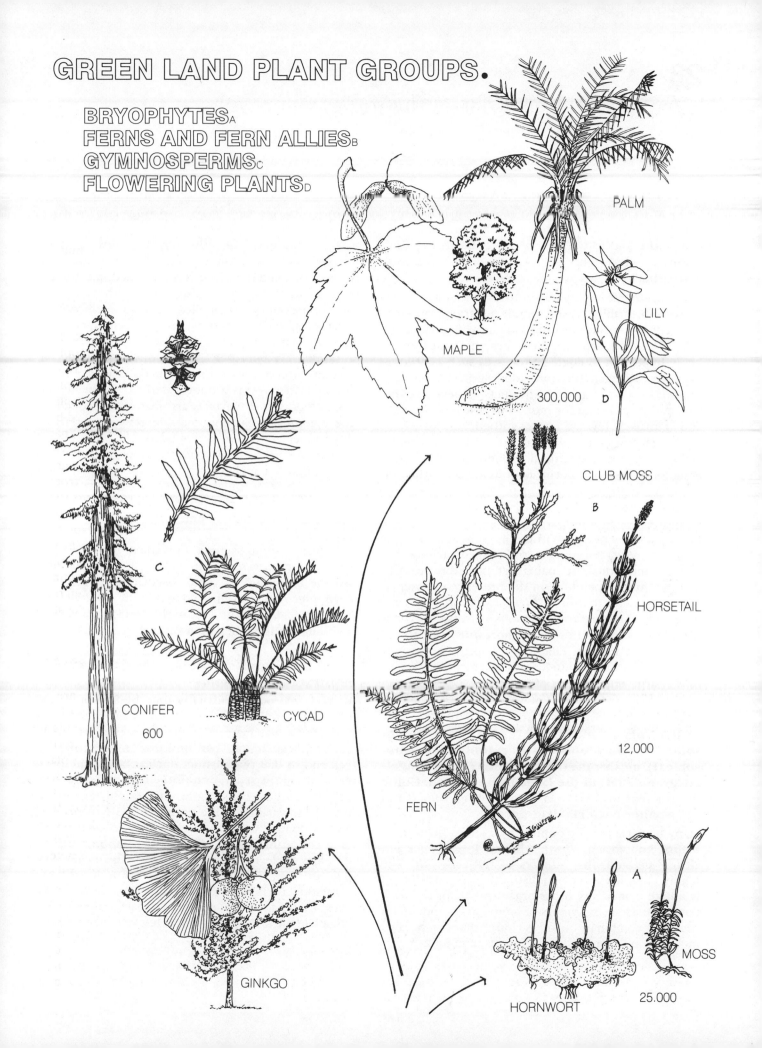

PALM

MAPLE

LILY

300,000   D

CLUB MOSS

B

HORSETAIL

C

CONIFER

600

CYCAD

12,000

FERN

GINKGO

HORNWORT

A

MOSS

25,000

23
BACTERIA (SCHIZOMYCOPHYTA)

Color the individual bacterial cells in the diagrams labeled "coccus" (A), "bacillus" (B), and "spirillus" (C).

Bacteria are minute organisms that are found in every conceivable habitat. A soil sample no larger than a pea may contain a hundred million individual bacterial cells. Most are saprophytic types that function in many important biological processes such as decomposition, nitrogen fixation, food spoilage, and the production of certain dairy products. Others are parasitic and responsible for numerous diseases of plants and animals. A few are photosynthetic.

Three major morphological types, based on overall cell shape, are recognized. Spherical cells, *cocci* (sing.: *coccus*), may occur as individuals, chains, or clusters (staphylococci), depending on species and environmental conditions. *Cocci* lack specialized structures for motility, and their movement appears to be rather random within the substrate. *Bacilli* (sing.: *bacillus*) are straight, rod-shaped bacteria that occur as individuals or chains that move directionally or randomly, depending on the presence or absence of structures for motility. *Spirilli* (sing.: *spirillus*) are long, helical-shaped rods that move directionally in a spirally twisting motion. Very short *spirilli* are sometimes called vibrios. Filaments, stalked forms, and other types are also found; and some form simple colonies (large aggregations of cells).

Color the gelatinous sheath or capsule (D), cell wall (E), plasma membrane (F), DNA or "chromosome" (G), nucleoplasm (H), cytoplasm (I), ribosome (J), mesosome (K), flagellum (L), and flagellar granule (M) in the large bacterial cell. Color the nuclear body (G + H) and the flagellum (L) in the smaller bacteria diagram.

Many bacteria are surrounded by a slimy, *gelatinous sheath,* known as a *capsule* when uniformly thick and clearly defined, that adheres to the *cell wall*. The highly complex and often distinctly layered *cell wall* is unlike that of any other organism. Major *cell wall* components are polymers of amino acids and amino sugars in combination and various other compounds including polysaccharides, proteins, and lipids.

The protoplast, delimited by the *plasma membrane,* is appressed to the *cell wall*. Bacterial cells are prokaryotic and therefore lack a defined, membrane-bound nucleus, nucleolus, and other double membrane-bound organelles, though membranous structures, such as photosynthetic membranes (not shown), may be present. A nuclear body consisting of a single *"chromosome,"* often one *DNA* molecule forming a continuous ring, imbedded in a *nucleoplasm* matrix forms a poorly defined region within the *cytoplasm*. *Ribosomes* are scattered throughout the *cytoplasm*. No endoplasmic reticulum (ER) or large aqueous vacuoles are present. The *mesosome,* a structure consisting of concentric or coiled membranes that is associated with the *plasma membrane* and thought to be functional in the separation of replicated *DNA* strands during cell division, is the only regularly membrane-bound structure within the *cytoplasm*.

In many motile bacteria, movement depends upon one or more *flagella* (whip-like organelles), that propel the bacterium by rapid, oscillating movements. Each *flagellum* is terminated within the *cytoplasm* by a small *flagellar granule*. Some bacteria are capable of a gliding motion when on a surface.

Color the diagrams at the bottom of the page that illustrate cell division.

Bacteria reproduce asexually only by fission. In the first stages of division, the cell elongates by the formation of new wall material in the mid region of the cell. Each replicate of the *"chromosome"* moves toward a cell end so that two distinct nuclear bodies are present. During the separation of the nuclear bodies, the *cell wall* begins constricting in the mid region and pinches inward from the periphery. The *mesosome,* which divides to produce one *mesosome* for each new cell, can be seen in close association with the dividing nuclear body. In unicellular forms, the two new cells separate from one another following the formation of new *cell walls* between the dividing cells. Under favorable conditions, bacterial cells may complete the division process in less than half an hour.

STRUCTURE AND REPRODUCTION.
BACTERIAL TYPES ★

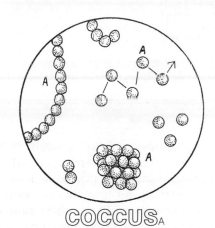

COCCUS_A

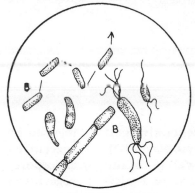

BACILLUS_B

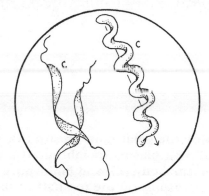

SPIRILLUS_C

BACTERIAL FISSION: ●
GELATINOUS SHEATH_D
CELL WALL_E
PROTOPLAST ★
   PLASMA MEMBRANE_F
   NUCLEAR BODY ★
     CHROMOSOME/DNA_G
     NUCLEOPLASM_H
CYTOPLASM_I
RIBOSOME_J
MESOSOME_K
FLAGELLUM_L
   FLAGELLAR GRANULE_M

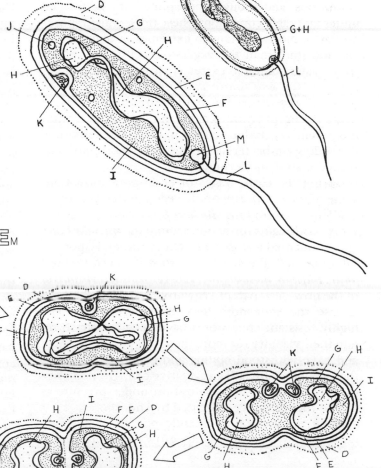

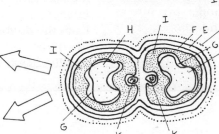

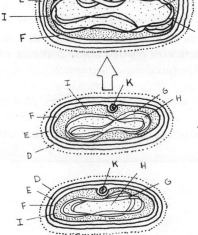

24

BACTERIA (SCHIZOMYCOPHYTA)

Color the gelatinous sheath (A), vegetative cell wall (B), plasma membrane (C), cytoplasm (D), nuclear body (E), and endospore wall (F) in the illustrations on the top half of the plate depicting endospore formation.

The major components of a vegetative bacterial cell are a *gelatinous sheath* (in many), *vegetative cell wall, plasma membrane, cytoplasm,* and *nuclear body* (disregarding the finer structures). Many bacteria form endospores under adverse environmental conditions, and though endospores may be formed under favorable conditions, these thick-walled spores are much more resistant to extremes of temperature and moisture than the vegetative cells. Also, endospore water content is extremely low, and the components of the *endospore wall* differ markedly from those of *vegetative cell walls.* Endospores vary in shape (from spherical to oval) and in position within the parent vegetative cell (from central to terminal), depending on bacterial species. The endospore illustrated is oval and terminal.

During the early stages of endospore formation, some of the *cytoplasm* of the vegetative cell becomes condensed around the *nuclear body.* Soon the *endospore wall* begins formation around the *nuclear body* and condensed *cytoplasm.* When the endospore is fully matured, the *gelatinous sheath, vegetative cell wall, plasma membrane,* and remaining *cytoplasm* of the now-dead parent vegetative cell disintegrate to release the endospore into the environment. The highly resistant endospores have been known to retain their viability for more than fifty years of storage and have withstood boiling and chemical treatment that would readily kill a vegetative cell. The necessity of conducting steam sterilization under pressure is due to the resistance exhibited by endospores.

Upon germination of an endospore, a single vegetative cell, which may rapidly grow to mature size, is released by each endospore. Since a single parent vegetative cell forms a single endospore and a single endospore, upon germination, produces a single vegetative cell, there is no population increase through endospore formation. The primary function of endospores is to carry a bacterium through periods of adverse environmental conditions and then reestablish it in the vegetative state when suitable conditions return.

Color the donor cell (G) and chromosome A (H) and the recipient cell (I) and chromosome B (J) in the illustrations on the bottom half of the plate depicting one form of genetic transfer in bacteria.

No true sexual reproduction is known to occur in bacteria, but various means of transferring genetic material, or hereditary information, between two bacterial cells have been observed. One means of genetic transfer (conjugation) occurs when one cell, the *donor cell,* donates a portion of its single chromosome to another cell, the *recipient cell,* which thereby gains additional genetic material. In the illustration, the *donor cell* contains *chromosome A* and the *recipient cell* contains *chromosome B.* Conjugation is initiated when a slender bridge, the conjugation tube (not colored) through which genetic material is transferred, forms between the two cells. Only a portion of the *donor cell's* chromosome, *chromosome A,* is usually transferred to the *recipient cell,* and the amount of genetic material transferred to the *recipient cell* apparently depends on how long the two cells remain in contact. The portion of *chromosome A* the *donor cell* donates is incorporated into *chromosome B* of the *recipient cell,* which then has hereditary information from the *donor* and usually exhibits some characteristics of the *donor cell* as well as its own. The *donor cell* usually dies as a result of the loss of genetic material.

ENDOSPORES AND GENETIC TRANSFER.

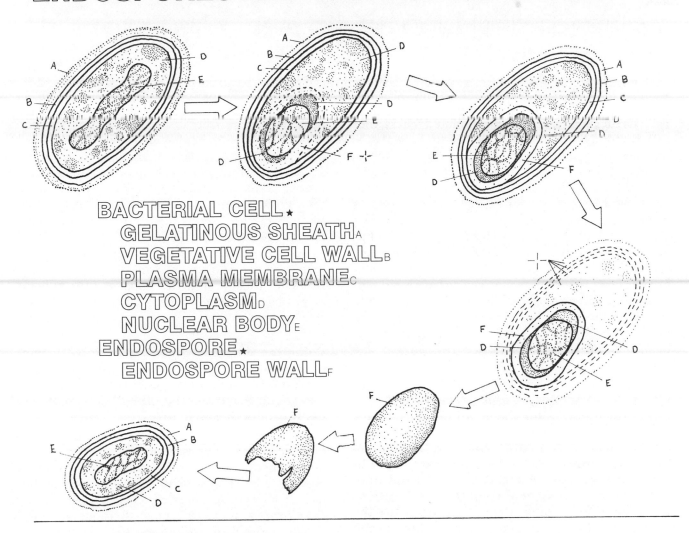

BACTERIAL CELL ★
   GELATINOUS SHEATH A
   VEGETATIVE CELL WALL B
   PLASMA MEMBRANE C
   CYTOPLASM D
   NUCLEAR BODY E
ENDOSPORE ★
   ENDOSPORE WALL F

GENETIC TRANSFER ★
   DONOR CELL G
   CHROMOSOME A H
   RECIPIENT CELL I
   CHROMOSOME B J

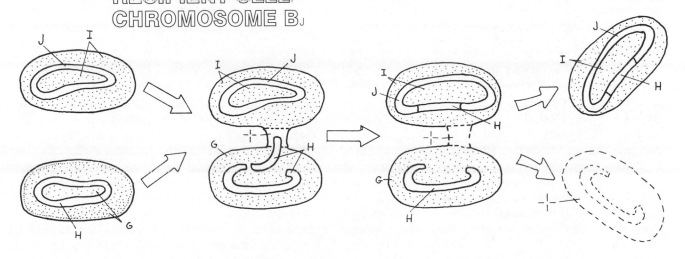

25

BLUE-GREEN ALGAE (CYANOPHYTA)

Blue-green algae, Cyanophyta, are widespread through a broad range of environments but are especially abundant in marine and continually moist habitats rich in nitrates. In some polluted waters, they may respond by producing a tremendous population increase, or "bloom," that often forms a wretched, stinking, floating mass that may smell of rotting garlic. One marine species, with dominant red pigments, produces red algal blooms and gave the Red Sea its name. On moist surfaces, such as overwatered porous clay plant pots, blue-green algae may form a blue-green to black slimy film or layer. Because many species fix atmospheric nitrogen into nitrogenous organic compounds, blue-green algae have some agricultural importance, especially in Asian rice paddies, where a blue-green alga, *Anabaena,* in association with an aquatic fern, *Azolla,* provide needed supplemental nitrates that enhance rice growth and production.

Color the examples of blue-green algae at the top of the plate. Color the gelatinous sheath (A), cell wall (B), plasma membrane (C), cytoplasm (D), nuclear region (E), photosynthetic membrane (F), ribosome (G), and gas vacuole (H) of the single cell and enlarged wedge. In addition, color the vegetative cells (I) of the two diagrams illustrating branching. Use a light color for the gelatinous sheath and cytoplasm and blue-green for the vegetative cells.

Like bacteria, blue-green algae are prokaryotic, and though little cellular differentiation is evident, they are physiologically complex. Though not as small as bacterial cells, individual cells are small, but many species are multicellular. Many unicellular, colonial, and filamentous species are surrounded by a secreted colorless *gelatinous sheath.* The blue-green algal *cell wall* is two-layered (not shown), with an outer mucilaginous pectic layer and an inner layer of stratified pectic and cellulosic compounds. The *plasma membrane,* or outer boundary of the protoplast, is appressed to the *cell wall.* As in bacteria, the *cytoplasm* contains no organized membrane-bound organelles. A *nuclear region,* consisting of nucleoplasm and DNA, is usually positioned centrally within the cell. All blue-green algae are photosynthetic, or autotrophic, and have *photosynthetic membranes* that are invaginations of the *plasma membrane,* but these are not surrounded by an enclosing membrane envelope as in eukaryotic cells. The prevailing color of blue-green algae is due to the presence of chlorophyll a (green) and phycocyanin (blue) pigments within the *photosynethic membranes. Ribosomes,* smaller than those of eukaryotic cells, are scattered throughout the *cytoplasm.* Some planktonic, or free-floating, species have *gas vacuoles* that apparently function in keeping them near the surface.

Two forms of branching are observed in filamentous blue-green algae. False branching occurs when a cell within a filament dies and one or both *vegetative cells* adjacent to the dead cell begin division to continue growth of the filament. The growing tip pushes through the *gelatinous sheath,* which continues to hold the filament together, to form a false branch because it is not due to a true lateral division. True branching occurs when a *vegetative cell* within the filament, usually near the tip, divides obliquely to produce a lateral growing tip that forms a true side branch.

Color the diagrams showing fission (J), endospores (K), exospores (L), akinete (M), heterocyst (N), hormogonia (O), separation discs (P), and vegetative cells (I) on the diagrams of asexual reproductive structures.

Sexual reproduction is currently unknown in the blue-green algae, but asexual reproduction, by various methods, is common. Most unicellular species reproduce by simple binary *fission,* but some form true aplanospores. Some attached species, such as *Dermocarpa,* undergo several divisions within the parent cell to form *endospores* that are released by rupture of the parent cell wall. Other attached unicells, such as *Chamaesiphon,* produce *exospores* by budding from the free end of the parent cell. Some filamentous species form *akinetes* from single *vegetative cells.* These enlarged, reserve food-rich cells have a thick wall and are usually highly resistant to environmental stress such as desiccation and temperature extremes. Upon germination, *akinetes* form new filaments.

Other filamentous species commonly produce *heterocysts,* modified *vegetative cells,* that function in nitrogen fixation and, due to fragmentation, may produce new filaments. Most filamentous species reproduce by simple fragmentation (not shown) as a result of natural breakage of the filament or by fragmentation involving specialized structures. *Hormogonia* are short segments of a filament that are formed when *vegetative cells* die and become *separation discs.* The filament breaks apart at the *separation discs* and releases the *hormogonia. Hormogonia* are produced under optimal environmental conditions and are not resistant. Each *hormogonium* continues growth as a new filament.

BLUE-GREEN ALGAE.

GELATINOUS SHEATH A
CELL WALL B
PROTOPLAST ★
  PLASMA MEMBRANE C
CYTOPLASM D
  NUCLEAR REGION E
  PHOTOSYNTHETIC
    MEMBRANE F
  RIBOSOME G
  GAS VACUOLE H
BRANCHING PATTERNS ★
VEGETATIVE CELL I

RICE PADDY

CLAY POT

AQUARIUM

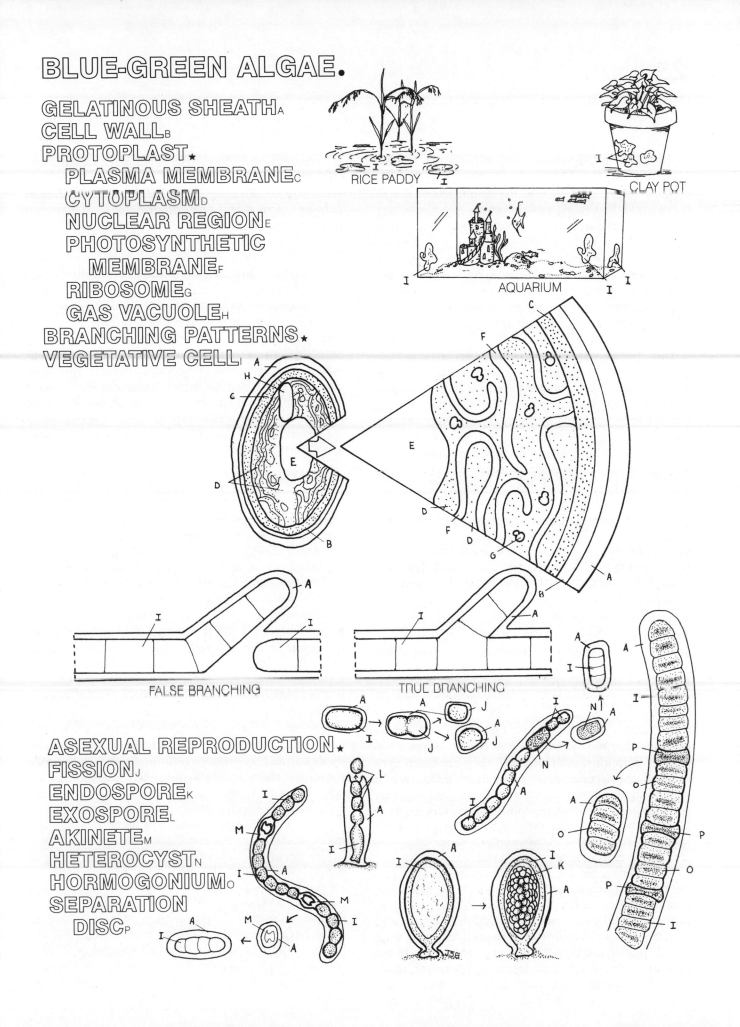

FALSE BRANCHING

TRUE BRANCHING

ASEXUAL REPRODUCTION ★
FISSION J
ENDOSPORE K
EXOSPORE L
AKINETE M
HETEROCYST N
HORMOGONIUM O
SEPARATION
  DISC P

Color the titles, arrow, and structures for the Myxomycophyta (A) and Myxomycetes (B) at the bottom of the plate. Use the same color for all structures in the Myxomycetes.

Five major classes in two subdivisions, *Myxomycophyta* (cellular and acellular slime molds) and *Eumycophyta* (true fungi), of fungi, division Mycota, can be recognized based on differences in vegetative and reproductive structures. Of the more than two hundred thousand known fungal species, no photosynthetic species exist, and most are saprophytic, though many are parasitic.

The *Myxomycophyta* (class *Myxomycetes*) are characterized by a vegetative stage consisting of a slimy protoplasmic mass capable of amoeboid movement. They lack cell walls, except around spores, and are multicellular by either an aggregation of naked cells or an acellular mass of multinucleate protoplasm. Both produce individual motile cells, either amoeboid or flagellated, at some stage in their life cycle. Only aplanospores, nonmotile spores, are produced. No hyphae, highly branched fungal filaments, are present. Slime molds usually inhabit moist, terrestrial habitats.

Color the titles, arrows, and structures for the Eumycophyta (C), Oomycetes (D), and Zygomycetes (E), using the title color for all structures within a group.

True fungi, *Eumycophyta,* produce rigid cell walls, and most produce a mycelium (an extensive hyphal mass) in their vegetative stage. Of the four major classes of true fungi, the *Oomycetes* (egg fungi—water molds, white rusts, and downy mildews) and *Zygomycetes* (zygote fungi—many common bread molds) are considered to be lower fungi because of their relatively simple structure. Most species in both classes produce hyphae lacking cross walls (septa) that delimit individual cells. Instead, they are coenocytic because the hyphal mass (mycelium) consists of a single, highly branched, multinucleate cell. Septa are present only between reproductive structures, both asexual and sexual, and the vegetative hyphae. A few species produce false, or incomplete, septa, and a few are motile or nonmotile unicellular species.

The *Oomycetes* are further characterized by production of biflagellated zoospores, with one tinsel and one whiplash flagellum, at some stage, primarily during asexual reproduction, of their life history. Cellulose is a major component of the cell wall, and vegetative hyphae are typically diploid. Following fusion in sexual reproduction, they produce dormant, thick-walled, resistant oospores containing one or more zygotes that form diploid hyphae upon oospore

germination. Both terrestrial and aquatic, marine and freshwater, species are found. Among the *Oomycetes* are many devastating plant parasites, some animal parasites, and many saprophytes. Two well-known plant parasites in this group are potato blight and grape downy mildew.

The *Zygomycetes* are a relatively small, mostly terrestrial group differing significantly from the *Oomycetes* in lacking zoospores and in having chitin as a major component of the cell wall. Aplanospores are typically produced within stalked sporangia elevated above the mycelial substrate. Hyphae are usually haploid in the vegetative stage and of two different mating strains. Fusion of compatible hyphae produces a zygote that develops into a thick-walled, resistant zygospore. Upon germination, zygospores divide meiotically to produce haploid aplanospores. Most *Zygomycetes* are saprophytic and account for substantial food losses.

Color the titles and structures for the Ascomycetes (F) and Basidiomycetes (G), using the same color for all structures within a group.

The *Ascomycetes* (sac fungi) and *Basidiomycetes* (club fungi) are considered to be higher fungi because they are relatively complex compared to the lower fungi. Most species in both classes produce hyphae having true crosswalls (septa) that delimit individual cells. Neither produces zoospores, and both have chitin as a major component of the cell wall. Hyphae are typically haploid in the vegetative stage and often binucleate, dikaryotic, during a portion of their sexual reproductive stage.

The *Ascomycetes* produce haploid aplanospores endogenously by meiosis within an enlarged, elongated cell, the ascus, formed terminally on a dikaryotic hypha. The four spores produced by meiosis usually undergo one synchronous mitotic division to produce a total of eight spores within each ascus. Asci are aggregated within a specialized structure, the ascocarp, in many species. Most species form mycelia, but some are unicellular. A number of plant parasites and many saprophytes are members of this group.

The *Basidiomycetes* resemble the *Ascomycetes* in many respects, but the club fungi produce aplanospores, called basidiospores, exogenously at the tip of a specialized sporangium, called a basidium, following meiosis within the sporangium. Many species produce large, complex fruiting bodies (mushrooms, puffballs, and shelf fungi), called basidiocarps, that consist of dense aggregations of dikaryotic hyphae. This group contains most of the fungi humans use as food.

INTRODUCTION TO THE FUNGI.

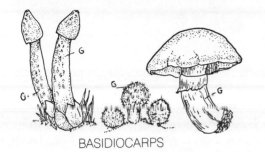

BASIDIOCARPS

SEPTATE
HYPHAE

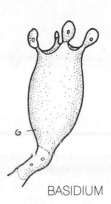

BASIDIUM

BASIDIOMYCETES G

APLANOSPORES

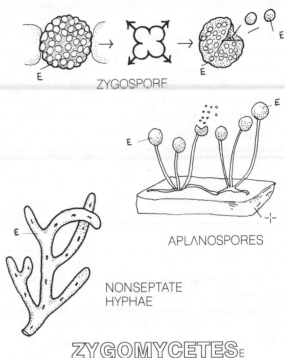

ZYGOSPORE

ASCOCARPS

SEPTATE
HYPHAE

ASCUS

E

NONSEPTATE
HYPHAE

APLANOSPORES

ASCOMYCETES F

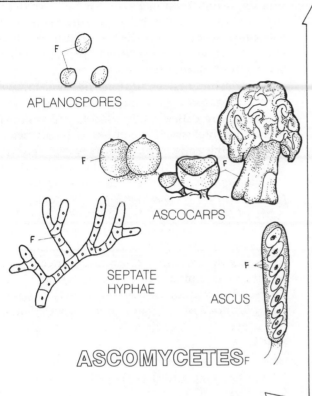

ZYGOMYCETES E

MYXOMYCOPHYTA A AND EUMYCOPHYTA C

A

C

OOMYCETES D

MYXOMYCETES B

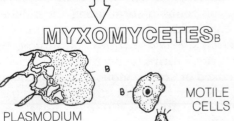

PLASMODIUM

MOTILE
CELLS

APLANOSPORE

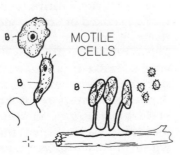

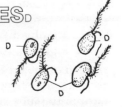

NONSEPTATE
HYPHAE

ZOOSPORES

OOSPORE

27
SLIME MOLDS (MYXOMYCETES)

Color the meiospores (A), swarm cell (B), myxamoebae (C), cyst (D), and gametes (E), starting at the top left of the plate. Stop coloring when you finish with the zygote (F) on the right.

The true (plasmodial) slime molds, class Myxomycetes, are a small group of about four hundred fifty widely distributed species. A number of life history variations exist, but *meiospores* of two different mating strains are typically produced. Under favorable conditions, they germinate to release a single wall-less cell, or two to eight cells if mitosis has occurred within the resting *meiospore*. *Meiospores* of slime molds are highly resistant and have been germinated after more than fifty years of storage. Two different types of motile cells, *swarm cells*, which typically have two whiplike anterior flagella for motility, and *myxamoeba*, which are motile by protoplasmic streaming, may be released, depending on species and environmental conditions. In species that produce either type, *swarm cells* are usually released if abundant water is present. Interconversion of the two types may occur, as indicated at the top of the plate. Under unfavorable environmental conditions, such as aridity, either of these motile haploid cells may form a thick-walled, resistant *cyst* that releases an active cell when conditions once again become favorable.

Active cells feed on bacteria and other organic materials. In some species, individual cells may undergo fission to produce additional haploid cells that will eventually function as *gametes*; in others, the cells released directly from the *meiospores* may soon function as *gametes*. In either case, syngamy of two haploid *gametes*, whether *swarm cells* or *myxamoebae*, form a wall-less *zygote*.

Color the plasmodia (G) and sclerotium (H) in the lower right quadrant.

Growth of the diploid generation is accompanied by a series of synchronous mitotic divisions that occur without the subsequent cytoplasmic division, or karyokinesis without cytokinesis, that would form individual cells. The result is called a *plasmodium*, a membrane-bound multinucleate protoplasmic mass capable of flowing, amoeboidlike movement due to protoplasmic streaming that is usually fan-shaped when moving from one point to another. An anastomosing network of veinlike tubules containing fluid cytoplasm, with the largest tubules toward the middle and trailing edge of the fan and smallest toward the leading edge where the *plasmodium* becomes a continuous mass of protoplasm, can be seen within the *plasmodium*. A slime trail, formed by remnants of the tubules, is left behind. The tubules are formed by protoplasm in a gel, or semisolid, state; and within the tubules, protoplasm in a solid, or fluid, state streams back and forth rhythmically, flowing in one direction for a few seconds and then reversing its flow (not shown). The greatest net flow of protoplasm is in the direction of plasmodium movement.

If environmental conditions become unfavorable for the *plasmodium*, it may form a crusty, dry, resistant structure called a *sclerotium*, which consists of numerous thick-walled, cell-like compartments containing living cells. A *plasmodium* will reform from the *sclerotium* when favorable conditions return.

A *zygote* may merge with another *zygote* or with a *plasmodium*; or a *plasmodium* may merge with another *plasmodium* (as shown in the lower right corner) to form an enlarged *plasmodium*, but most *plasmodium* growth is due to cytoplasmic increase by metabolic processes. *Plasmodia* range in size from less than a centimeter to over a meter in diameter. At times, a fragment of a *plasmodium* may be left behind as a *plasmodium* migrates (not shown). These fragments are capable of development as a separate *plasmodium*.

Color the hypothallus (J), stipe (K), and meiosporangium (L) of the fructifications (I) on the left. Using the fructification (I) color, color over the whole fructification-forming plasmodium in the lower left corner.

Under favorable environmental conditions, a *plasmodium* will migrate to an exposed, usually elevated, surface. When the *plasmodium* settles, masses of protoplasm form along the tubules and develop into meiospore-forming structures called *fructifications*. *Fructifications*, or fruiting bodies, occur in various forms, depending on species. They may be stalked or sessile, globose to elongate or relatively amorphous. Each *fructification* has a base, the *hypothallus*, that attaches it to the substrate surface, and if stalked, a *stipe* that elevates the *meiosporangium* above the surface to facilitate dispersal of released *meiospores* by air currents. The wall of the *meiosporangium* surrounding the meiospore-forming cells is called the perideridium. Meiosis within the *meiosporangium* produces *meiospores*, which are released when the fragile perideridium splits or breaks apart.

SLIME MOLD LIFE HISTORY.

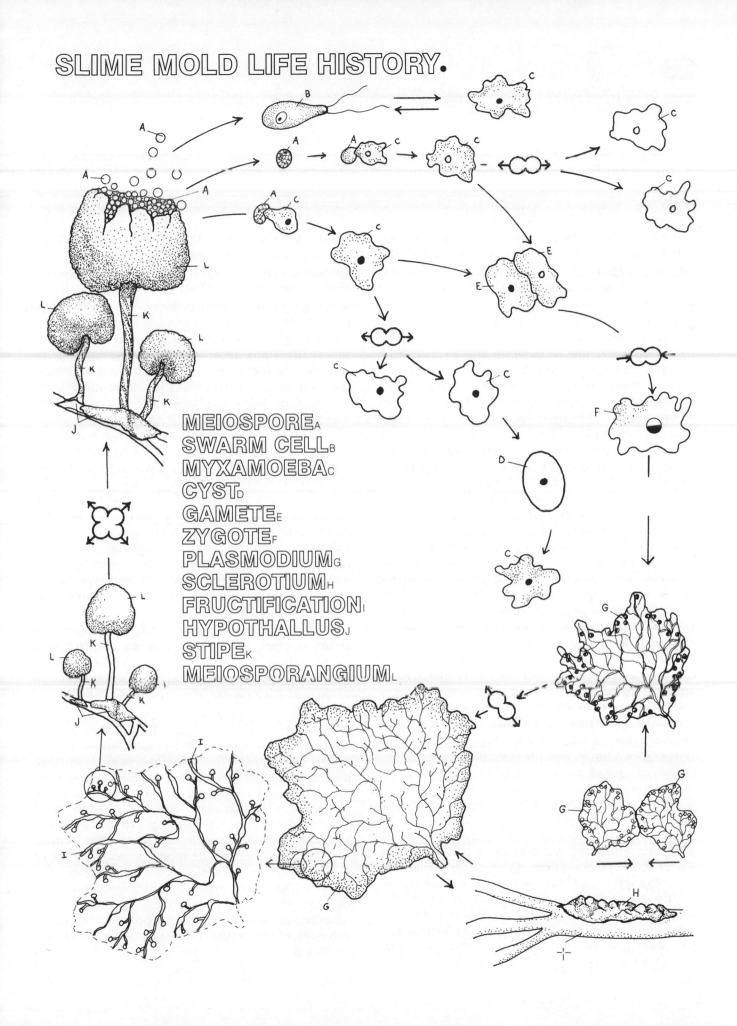

MEIOSPORE A
SWARM CELL B
MYXAMOEBA C
CYST D
GAMETE E
ZYGOTE F
PLASMODIUM G
SCLEROTIUM H
FRUCTIFICATION I
HYPOTHALLUS J
STIPE K
MEIOSPORANGIUM L

28

EGG FUNGI (OOMYCETES)

Color the nonseptate hyphae (A), septa (B), antheridium (C), sperm nuclei (D), oogonium (E), egg (F), zygote (G), and oospore (H) on the diagrams depicting sexual reproduction on the top half of the plate.

The Oomycetes (egg fungi) typically produce a dense vegetative mycelium composed of a multinucleate, complexly branched, *nonseptate hyphae* mass. Both sexual and asexual reproduction occur, and reproductive structures of both types are separated from the vegetative *nonseptate hyphae* by distinct *septa,* or crosswalls. The life cycle of a water mold, *Saprolegnia,* provides an example of the general life cycle and morphological features of an Oomycetes fungus.

Sexual reproduction in *Saprolegnia* usually begins when vegetative *nonseptate hyphae* of two mating strains, male and female, grow adjacent to one another. A chemical message the female strain secretes induces *antheridia,* or male gametangia, formation on the male strain, which responds by secreting a corresponding chemical message inducing *oogonia,* or female gametangia, formation on the female strain. Haploid nuclei are formed within each type of gametangium by meiosis. Depending on species, *antheridia* and *oogonia* contain one to numerous haploid nuclei following meiosis. *Sperm* consist of a single haploid male nucleus; *eggs* consist of a single haploid female nucleus surrounded by ample cytoplasm.

On contact with the *oogonium,* an *antheridium* clasps the *oogonium* and produces one or more conjugation tubes that penetrate the *oogonium* and release *sperm nuclei* into the oogonial chamber. Upon fusion of *sperm* with the *egg* or *eggs* within an *oogonium,* one or more *zygotes* are formed. A thick-walled, resistant *oospore* that surrounds the *zygotes* forms within the *oogonium.* Release of the *oospore* occurs through disintegration of the oogonial cell wall, but it remains dormant for a period of time. Under favorable conditions, the *oospore* germinates, through rupture of the oospore wall, and the *zygotes* within grow into diploid, vegetative, *nonseptate hyphae.* These may continue growth as *nonseptate hyphae* and eventually undergo sexual reproduction, or, under low nutrient conditions, they may form asexual structures.

Color the sporangia (I), asexual zoospores (J), resting spores (J¹), and the previously listed structures on the diagrams of asexual reproduction across the lower half of the plate.

Sporangium development begins as an enlarged hyphal tip containing cytoplasm and numerous diploid nuclei. A *septum* that separates the developing *sporangium* from the vegetative *nonseptate hyphae* soon forms. Portions of cytoplasm surrounding each nucleus then condense and produce a cell wall to form numerous developing diploid primary *asexual zoospores.* Eventually, an opening forms in the tip of the *sporangium* through which the mature primary *asexual zoospores,* each with two anterior flagella, are released. The anteriorly flagellated primary *asexual zoospores* are motile for a period, but they eventually lose their flagella and form thick-walled, resistant *resting spores* that remain dormant during unfavorable environmental conditions. When favorable conditions return, the *resting spore* germinates and releases a single, laterally biflagellated secondary *asexual zoospore* that develops into a vegetative *nonseptate hyphae* on a suitable substrate.

Color the sporangiophore (K) and associated structures previously listed on the potato blight diagram in the lower left corner of the plate.

Terrestrial *asexual zoospore* formation is exemplified by the potato blight fungus, *Phytophthora,* which forms a dense mycelium of *nonseptate hyphae* within the tissues of a potato leaf. The *hyphae* secrete enzymes that digest the leaf tissues for use by the fungus. At maturity, *sporangiophores* grow through the leaf's stomatal pores and form *sporangia* at their tips. The small *sporangia* do not release *asexual zoospores* while attached, but are releasd intact and dispersed by wind. If *sporangia* land on a suitable substrate, a potato leaf in this case, they remain quiescent until damp conditions provide a film of water on the leaf's surface. Dampness induces the release of *asexual zoospores,* which then swim to a site suitable for germination near a stomatal pore. Germinating *asexual zoospores* produce vegetative *nonseptate hyphae* that enter stomatal pores to infect the leaf.

EGG FUNGI LIFE HISTORY.

SEXUAL REPRODUCTION ★
ANTHERIDIUMc
SPERM NUCLEUSd
OOGONIUMe
EGGf
ZYGOTEg
OOSPOREh

NONSEPTATE HYPHAa
SEPTUMb

POTATO LEAF

ASEXUAL REPRODUCTION ★
SPORANGIUMi
ASEXUAL ZOOSPOREj
RESTING SPOREj'
SPORANGIOPHOREk

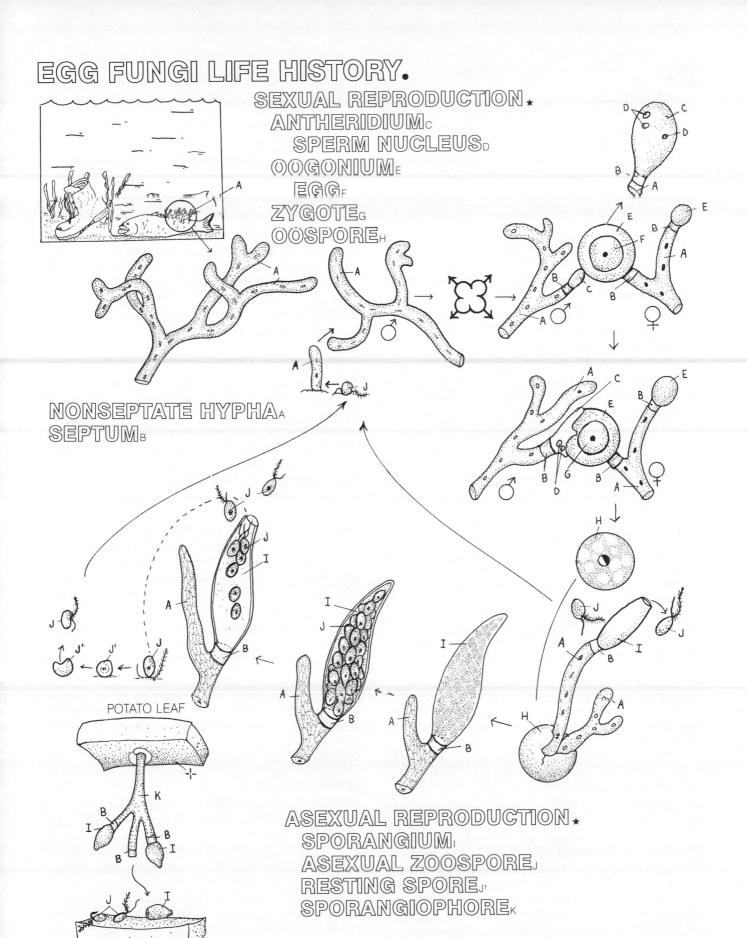

29
ZYGOTE FUNGI (ZYGOMYCETES)

Color the meiospores (A) and hyphae (B) starting at the top center of the plate and stopping when you finish with the branched hyphae below the first bread slice.

Common black bread mold, *Rhizopus stolonifera,* provides an example of a Zygomycetes life history. In black bread mold, a saprophytic fungus responsible for substantial food losses, lightweight haploid *meiospores,* produced by sexual reproduction, may be dispersed over long distances by air currents. *Meiospores* are ubiquitous, and those that land on a suitable substrate, such as a moist piece of unpreserved bread or soft fruit, readily germinate and form an extensive, nutrient-gathering, vegetative mycelium consisting of delicate, nonseptate, branched, tubelike *hyphae* containing numerous haploid nuclei. The mycelium permeates and spreads over the surface of the substrate to form a soft, whitish-gray, cottony mass. Black bread mold *hyphae* secrete exogenous digestive enzymes that break down the substrate to release nutrients the *hyphae* can absorb.

Color the stolons (C), rhizoids (D), asexual sporangiophores (E), mitosporangia (F), and mitospores (G) under the "asexual reproduction" label. Though (C), (D), and (E) are continuous tubes without septa to separate them, they are illustrated as separate structures for clarity and coloring purposes.

Asexual reproduction, which may be initiated soon after a mycelium is formed, begins with the formation of specialized *hyphae,* called *stolons. Stolons,* which are thicker than the vegetative *hyphae,* arch across the substrate surface and anchor to it at intervals by short, branching, rootlike *hyphae,* called *rhizoids,* that penetrate the substrate. At each anchor point, clusters of stalklike *asexual sporangiophores,* terminated by an enlarged tip, develop. The enlarged tip becomes separated from the *asexual sporangiophore* by a septum and forms an asexual sporangium, or *mitosporangium.* Numerous mitotic divisions within each *mitosporangium* produce a multitude of genetically identical, lightweight, haploid *mitospores* that are released by rupture of the mitosporangial wall.

Like meiospores, *mitospores* function to proliferate and distribute the black bread mold fungus. Though two different mating strains, plus and minus since they are homomorphic, are required for sexual reproduction, a single mating strain may be rapidly proliferated and widely distributed through asexual reproduction.

Color progametangia (H), gametangia (I), suspensors (J), zygote (K), zygospore (L), sexual sporangiophore (M), meiosporangium (N), and previously listed structures on the diagrams illustrating sexual reproduction up the left margin.

Unless mycelia of both mating strains grow within the same substrate, reproduction is strictly asexual. Sexual reproduction is initiated when two *hyphae* of compatible mating strains grow in close proximity and exchange chemical messages, or sex hormones, that induce the formation of *progametangia.* Each *hypha,* plus and minus, forms a pouchlike outgrowth, or *progametangium,* which grow toward each other. When the two *progametangia* contact one another, each produces a septum that cuts off a multinucleate *gametangium* and a supporting *suspensor.* The walls of the plus and minus *gametangia* soon dissolve at their region of contact to form a single multinucleate cell containing paired haploid nuclei of both mating strains. The two nuclei of each pair then fuse to form a multinucleate *zygote* containing numerous diploid nuclei. The *zygote* then produces a thick, resistant spore wall to form a dormant *zygospore* that functions to provide limited dispersal and to carry the black bread mold through periods of adverse environmental conditions.

When conditions favorable for black bread mold growth recur, the numerous diploid nuclei within the *zygospore* undergo meiosis to produce haploid nuclei of both mating strains. The *zygospore* then germinates by rupture of its thick wall, and a single *sexual sporangiophore* terminated by a *meiosporangium* containing numerous haploid *meiospores* of both mating strains is produced. Release of mature *meiospores* is through rupture of the meiosporangial wall.

BLACK BREAD MOLD LIFE HISTORY.

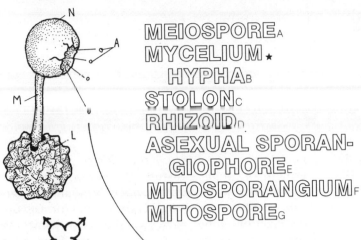

MEIOSPOREA
MYCELIUM ★
   HYPHAB
STOLONC
RHIZOIDD
ASEXUAL SPORAN-
   GIOPHOREE
MITOSPORANGIUMF
MITOSPOREG

PROGAMETANGIUMH
GAMETANGIUMI
SUSPENSORJ
ZYGOTEK
ZYGOSPOREL
SEXUAL SPORANGIO-
   PHOREM
MEIOSPORANGIUMN

SEXUAL REPRODUCTION ★

ASEXUAL REPRODUCTION ★

UNINFECTED
BREAD ★

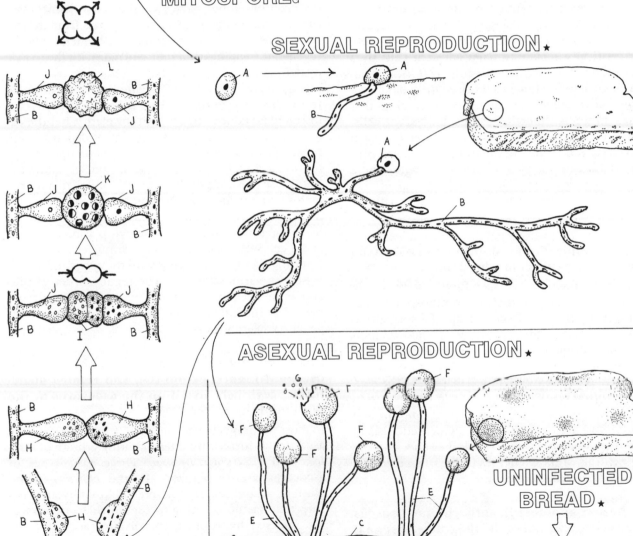

30

SAC FUNGI (ASCOMYCETES)

Color the mycelium (A), haploid hypha (A¹), dikaryotic hypha (B), cell wall (C), and septum (D) at the top of the plate. Color the morel mushroom as well.

The Ascomycetes (sac fungi) produce an extensive *mycelium* consisting of highly branched, septate hyphae. In vegetative *haploid hyphae,* each cell, characteristically separated from adjacent cells by a perforated *septum* that is part of the *cell wall,* contains a single haploid nucleus. Each *septum* has a central *pore,* surrounded by a thickened septal ring, that provides an interconnection between adjacent cells. *Dikaryotic hyphae,* which have two nuclei in each cell (usually one from each mating strain), are produced by fusion of two *haploid hyphae.* These binucleate hyphae may continue to grow vegetatively for long periods and may also reproduce asexually by the formation of dikaryotic spores (not shown).

Color the conidia (F) and conidiophore (G) on the diagrams of asexual reproduction.

Asexual reproduction in sac fungi is primarily through formation of *conidia,* or conidiospores, that are continually pinched from the tip of a specialized hypha, the *conidiophore,* by mitotic division. As indicated by the numbered sequence on the diagrams, the most distal conidiospore is the oldest. *Conidia* remain attached to one another until mature, so the *conidiophore* often bears a beadlike chain of *conidia* in various stages of maturation. Masses of *conidia* and *conidiophores* are evident as the pigmented areas in Roquefort, Camembert, and some other cheeses. The slight grittiness of these cheeses is due to *conidia* being crushed between the teeth.

Color the ascogonium (H), antheridium (I), ascus (J), and ascospores (K) in the diagrams of sexual reproduction.

Sexual reproduction in the Ascomycetes begins when two *haploid hyphae* of different mating strains grow toward one another in the substrate and gametangia start forming. The female strain produces a multinucleate *ascogonium,* a special type of oogonium that contains numerous female nuclei. The male strain produces an *antheridium* containing numerous male nuclei. The *ascogonium* then forms a tubular outgrowth, or conjugation tube, that contacts the *antheridium.* The walls of the *antheridium* and *ascogonium* dissolve at the point of contact and form an open conjugation tube between the two gametangia through which the male nuclei may pass from the *antheridium* to the *ascogonium.* Male nuclei migrate through the conjugation tube and pair with the female nuclei.

Binucleate, *dikaryotic hyphae* are soon initiated from the *ascogonium.* The *dikaryotic hypha* illustrated is reduced to one cell for clarity, but the extent of dikaryotic hyphal development depends upon species. One or more *dikaryotic hyphae* tips enlarge to form a binucleate *ascus.* The two nuclei within the *ascus* fuse to form a diploid nucleus and then undergo meiosis to produce four haploid nuclei. These nuclei usually undergo a synchronous mitotic division to form eight haploid nuclei. The cytoplasm condenses and a cell wall forms around each nucleus; the eight resulting *ascospores* mature with the *ascus* and are released by various means. The regular order of plus and minus strain *ascospores* within the *ascus* has made some members, such as *Neurospora,* important genetic research tools.

Color the cleistothecium (L), perithecium (M), ostiole (N), apothecium (O), and related structures previously listed on the ascocarps at the bottom of the plate.

In most Ascomycetes, a complex "fruiting body," the ascocarp, is formed by interdigitating growth of *haploid* and *dikaryotic hyphae.* Three basic types of ascocarps are recognized. In the *cleistothecium,* clusters of *asci* are completely enclosed by an ascocarp that has no opening to the outside. *Ascospores* are released by rupture, or disintegration, of the *cleistothecium. Perithecia* surround the clusters of *asci,* but a pore, the *ostiole,* provides a connection with the outside through which *ascospores* may be released. *Apothecia* are open, cup-shaped ascocarps in which the *asci* are arranged on an exposed surface. The morel mushroom is an aggregation of *apothecia* supported by a stalk. Some species produce naked *asci* individually or in clusters that are not imbedded in an ascocarp, as illustrated on the far right.

SAC FUNGI · INTRODUCTION.

MOREL

MYCELIUM A
HYPHA ★
HAPLOID A'
DIKARYOTIC B
CELL WALL C
SEPTUM D
PORE E

ASEXUAL
REPRODUCTION ★
CONIDIUM F
CONIDIOPHORE G

BLUE CHEESE

SEXUAL REPRODUCTION ★
ASCOGONIUM H
ANTHERIDIUM I
ASCUS J
ASCOSPORE K

ASCOCARPS ★
CLEISTOTHECIUM L
PERITHECIUM M
OSTIOLE N
APOTHECIUM O

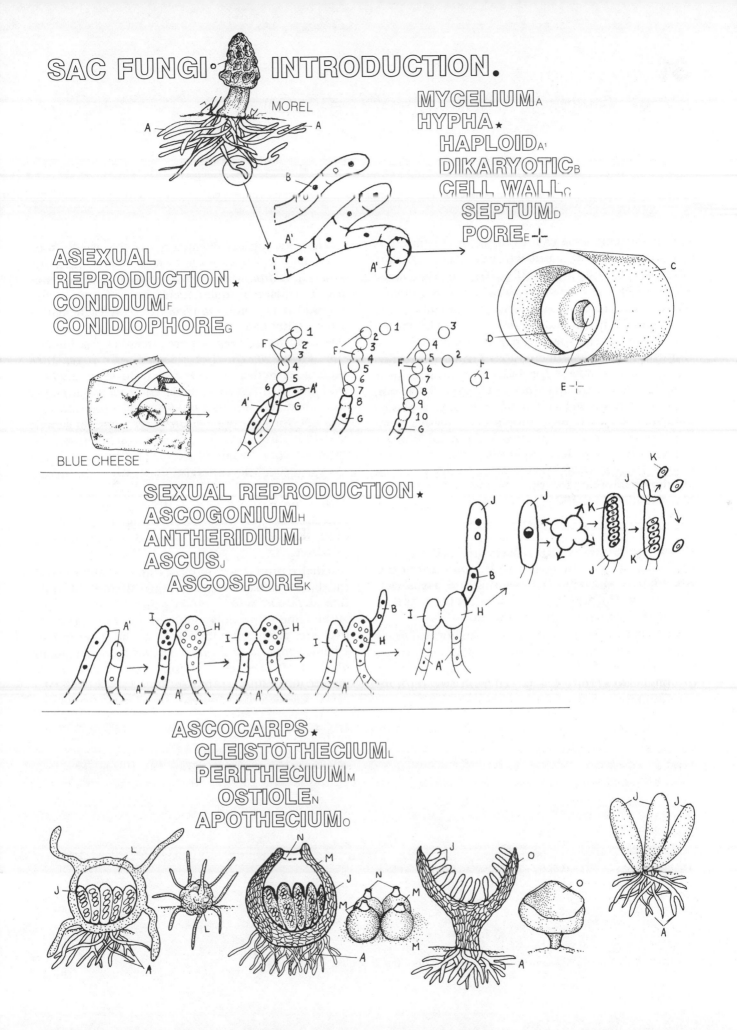

31
SAC FUNGI (ASCOMYCETES)

Color the cell wall (A), cytoplasm (B), nucleus (C), nucleolus (D), mitochondria (E), storage granules (F), and vacuole (G) on the large diagram of a yeast cell at the top of the plate.

Many yeasts, class Ascomycetes, are ubiquitous sac fungi widely distributed by air currents. The natural occurrence of yeast species on fruit surfaces is responsible for the natural fermentation of fruits and the formation of sourdough for baking. Typical cellular structure, as illustrated in the large yeast cell diagram, includes a nonliving *cell wall*, enclosing a plasma membrane-bound living protoplast. The plasma membrane is not shown. Contents of the *cytoplasm* include a *nucleus* with a single *nucleolus, mitochondria, storage granules,* and a large *vacuole.* Yeasts are diplohaplontic; and though isogamous, they are heterothallic since the gametes must be of two different, plus and minus, mating strains.

Color the gametes (H), dikaryotic cell (I), and diploid cell (J) in the upper right quadrant of the life history diagram. Also color the ascus (K), ascospores (L), and haploid cells (M) to complete the sexual reproduction diagram.

Sexual reproduction involves the fusion of two *gametes* that come together and fuse without occurrence of karyogamy (fusion of the two *nuclei*), thereby forming a *dikaryotic cell* (a cell with two separate *nuclei*). Karyogamy eventually occurs, usually soon after the dikaryotic cell is formed, to produce a *diploid cell.* A *diploid cell* that is going to undergo meiosis in sexual reproduction enlarges to form a saclike *ascus.* Each haploid *nucleus,* two of each mating strain, within an *ascus* forms an *ascospore* to produce four *ascospores* that are released by disintegration of the *ascus.* The released *ascospores* then undergo growth and maturation to form mature *haploid cells* that, under suitable environmental conditions, may function as *gametes.* Sexual reproduction accounts for some population increase and for genetic variation through meiosis and recombination. The most effective means of population increase, however, is through asexual reproduction.

Color the haploid and diploid parent cells (N) and the buds (O) they are forming, as well as the haploid cells (M) and diploid cells (J) they produce.

Asexual reproduction in yeasts may occur in either the diploid or haploid generations. Mature cells function as *parent cells* in either generation and undergo budding to produce diminutive offsets, or *buds,* that are genetically identical to the *parent cell,* whether *haploid* or *diploid.* The *buds* are eventually pinched off and released from the *parent cell* to mature into adults.

YEAST LIFE HISTORY.

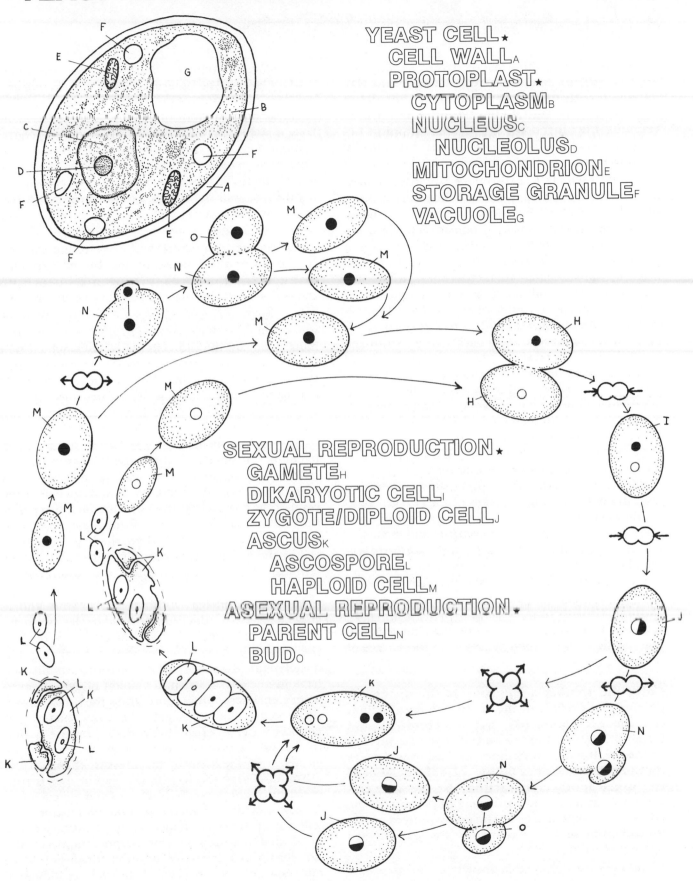

YEAST CELL.★
CELL WALL A
PROTOPLAST.★
CYTOPLASM B
NUCLEUS C
NUCLEOLUS D
MITOCHONDRION E
STORAGE GRANULE F
VACUOLE G

SEXUAL REPRODUCTION.★
GAMETE H
DIKARYOTIC CELL I
ZYGOTE/DIPLOID CELL J
ASCUS K
ASCOSPORE L
HAPLOID CELL M
ASEXUAL REPRODUCTION.★
PARENT CELL N
BUD O

32
SAC FUNGI (ASCOMYCETES)

Color the germinating ascospores (A), haploid hyphae (B), antheridium (C), ascogonium (D), and dikaryotic hyphae (E), starting at the top left below the diagram labeled "apothecia," which remains uncolored for now, and stopping below the sectioned apothecium at the middle right margin.

Cup fungi, such as *Peziza,* produce a multitude of dry, lightweight *ascospores,* which are the primary means of dispersal and proliferation. *Ascospores* are meiospores and are either male or female. When a wind-dispersed *ascospore* lands on a suitable substrate, it germinates and produces a mycelium, consisting of a mass of *haploid hyphae* with crosswalls, that permeates the substrate and functions in nutrient uptake.

In cup fungi, the formation of male and female gametangia, *antheridium* and *ascogonium,* respectively, is induced by an exchange of chemical messages, through sex hormones, between male and female *haploid hyphae* growing in close proximity within the same substrate. *Antheridia* and *ascogonia,* which develop as terminal enlargements on *haploid hyphae,* are both multinucleate, with numerous haploid nuclei. The larger *ascogonium* soon forms a tubular outgrowth, the conjugation tube (not separately colored), that grows toward the smaller *antheridium.*

When the conjugation tube of the *ascogonium* contacts the *antheridium,* the walls of the two gametangia dissolve at the point of contact to form an open passageway between the *antheridium* and *ascogonium.* The haploid male nuclei from the *antheridium* then migrate through the open conjugation tube into the *ascogonium.* Each male nucleus then pairs, but does not fuse, with a female nucleus within the *ascogonium.* Growth of *dikaryotic hyphae,* with crosswalls, in which each cell has one male nucleus and one female nucleus, soon begins from the surface of the *ascogonium.*

Color the apothecia (F) group in the upper left corner; the surface of the sectioned apothecium (F); the excipulum (G), hypothecium (H), and hymenium (I) within the boxed area of a portion of the sectioned apothecium; and all three layers on the orientation bar below the enlargement of the boxed apothecium section. Use the same color for (B) and (G) and a mixture of (B) and (E) for (H). Color the features of the enlargement of the apothecium section except for the paraphyses (J) and asci (K).

The *dikaryotic hyphae* that grow from the *ascogonium* do not permeate the substrate as a nutrient-gathering mycelium. Instead, the *dikaryotic hyphae,* as well as *haploid hyphae,* begin to develop an ascocarp or "fruiting body," which in cup fungi is of

the *apothecium* type. An *apothecium* has three intergrading layers. The outermost layer, the *excipulum,* covering the outside of the "cup," is a protective layer consisting of densely packed *haploid hyphae.* This layer is also called the peridium. The middle layer, the *hypothecium,* consists of both *haploid hyphae* and *dikaryotic hyphae* that are closely intermingled with one another. The innermost layer, the *hymenium,* lining the inside of the cup, is a fertile layer that forms the lining of the cup.

Color the paraphyses (J) and developing asci (K) in the hymenium layer of the boxed enlargement. Color the ascus-formation sequence up the left side of the plate. Use the (B) color for (J).

The fertile *hymenium* consists of a mixture of elongate terminal cells of *haploid hyphae* and larger terminal elongate diploid cells called *asci,* which will undergo meiosis to produce *ascospores. Ascus* formation begins with a terminal dikaryotic cell (figure 1) that bends to form a hook-shaped structure called a crozier (figure 2). Mitotic division of both haploid nuclei, male and female, within the crozier cell produces a cell with four nuclei (figure 3). Cell walls then form to cut off one nucleus at the crozier tip and one at the crozier base (figure 4), leaving a large dikaryotic cell with one male nucleus and one female nucleus at the bend of the crozier (figure 4). The uninucleate crozier tip cell and base cell fuse to form a dikaryotic cell (figure 5). The two nuclei within the large dikaryotic cell at the bend of the crozier fuse to form a large terminal diploid cell (zygote), the young *ascus* (figure 6).

The diploid nucleus within the *ascus* soon undergoes meiosis to produce four haploid nuclei, two male and two female (figure 7). A synchronous mitotic division of all four haploid nuclei produces eight haploid nuclei (figure 8). A spore wall soon forms around each haploid nucleus and a small amount of cytoplasm to form eight *ascospores* (figure 9). As the *ascus,* now a single cell containing eight *ascospores,* matures, it takes in water to create hydrostatic pressure within the *ascus.* At maturity, the *ascospores* are forcibly ejected from the *ascus* by this pressure (figure 10).

The discharge mechanism on mature *asci* is triggered by slight pressure, such as from air currents. Gently blowing into an *apothecium* with mature *asci* will trigger the ejection mechanism, creating a cloud of ejected *ascospores.* This adaptation places the *ascospores* in a favorable position for dispersal. In most species, no specialized asexual reproductive structures are known. In all species, asexual reproduction may occur through fragmentation of the haploid hyphae mycelium.

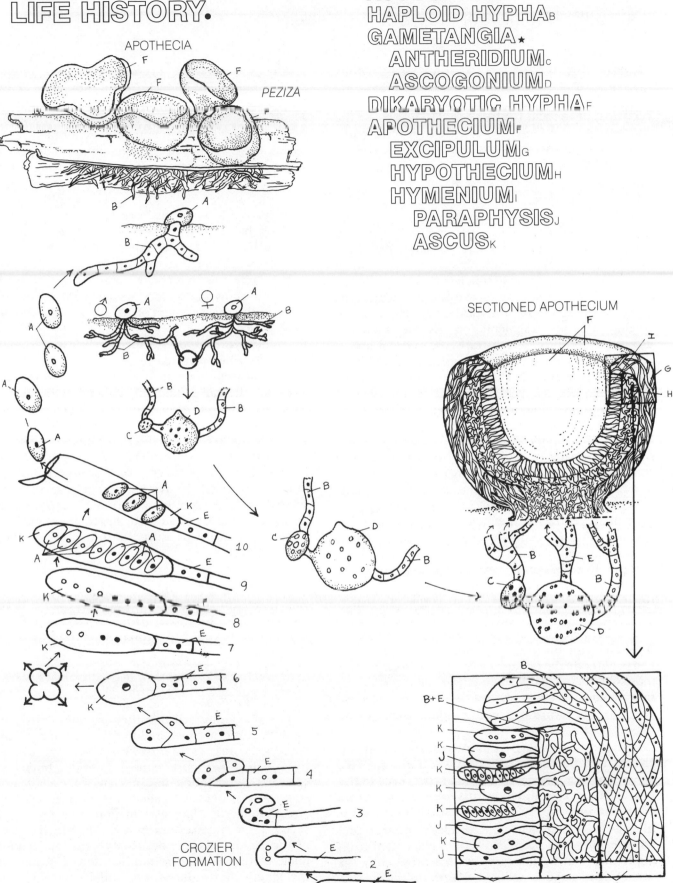

CUP FUNGUS LIFE HISTORY.

APOTHECIA

PEZIZA

ASCOSPORE A
HAPLOID HYPHA B
GAMETANGIA ★
ANTHERIDIUM C
ASCOGONIUM D
DIKARYOTIC HYPHA E
APOTHECIUM F
EXCIPULUM G
HYPOTHECIUM H
HYMENIUM I
PARAPHYSIS J
ASCUS K

SECTIONED APOTHECIUM

CROZIER
FORMATION

33
CLUB FUNGI (BASIDIOMYCETES)

Color the haploid hyphae (A), dikaryotic hyphae (B), and spores (C). Note that the mycelium below the mushroom at the upper left is a combination of (A) and (B) hyphae. Do not color the illustrations labeled "sexual" yet, but do color the two diagrams illustrating mushrooms (F) on the top half of the plate.

The club or fleshy fungi, class Basidiomycetes, produce an extensive, nutrient-gathering, vegetative *mycelium* within a substrate consisting of *haploid hyphae, dikaryotic hyphae,* or a combination of both. Asexual reproduction is primarily through fragmentation in subclass Homobasidiomycetidae and through asexual *spore* formation in subclass Heterobasidiomycetidae. The typical growth pattern of a developing subsurface *mycelium,* outward from its point of origin, promotes fragmentation. As the *mycelium* grows farther into the substrate, the nutrient materials available in the substrate occupied by older *mycelium* become depleted. The older portions of the *mycelium* may die as a result, leaving a vacant center. A continually enlarging subsurface ring of *mycelium,* or "fairy ring," may develop. The *mycelium,* which may be up to a few hundred years old, is evident only seasonally as a ring of *mushrooms.* In a homogenous substrate, the integrity of the ring may be maintained indefinitely, but a heterogenous substrate or disturbance often disrupts the continuity of mycelial growth, resulting in fragmentation. Each mycelial fragment thus formed may then continue growth as an independent plant.

Many Heterobasidiomycetidae reproduce asexually by sporulation, which provides an efficient adaptation for long-distance dispersal and rapid population increase. Enormous numbers of *spores* must be produced because they are wind dispersed, and only those that land on an appropriate host will germinate and proliferate the fungus. Asexual *spore* formation in most species occurs through continual mitosis at the tips of specialized hyphae in a manner similar to conidiospore formation found in some Ascomycetes.

Color the basidium (D), basidiospores (E), mushrooms (F), puffballs (G), and shelf fungus (H), as well as the previously listed structures in the top half of the illustrations depicting sexual reproduction.

In subclass Homobasidiomycetidae, *basidia* and *basidiospores* are produced within elaborate "fruiting bodies," or basidiocarps, which develop by the dense, aggregated growth of *dikaryotic hyphae.* In this group, two *haploid hyphae* of compatible mating strains fuse to form a binucleate cell that then produces *dikaryotic hyphae.* The two *haploid hyphae* must be of opposite mating strains in heterothallic species; in homothallic species, the *haploid hyphae* may be of the same mating strain. *Dikaryotic hyphae* may grow indefinitely without forming basidiocarps, but when environmental conditions are favorable, basidiocarps are produced. Within basidiocarps, numerous terminal dikaryotic cells enlarge to form club-shaped *basidia.* The two haploid nuclei in each dikaryotic *basidium* fuse to form a single diploid nucleus that undergoes meiosis to produce four haploid nuclei, two of each mating strain in heterothallic species, which migrate toward the tip of the *basidium,* where four pouchlike outgrowths, the sterigmata (not shown), develop. A single haploid nucleus, along with a small amount of cytoplasm, migrates into each pouch, where each produces a spore wall to form four *basidiospores,* which are released from the *basidium* when mature.

Basidiocarps come in many different forms and sizes, but the most commonly encountered types are the *mushroom, puffball,* and *shelf* or bracket *fungus.* *Mushrooms* produce *basidiospores* within a caplike structure that terminates in a stalk and freely disperse *basidiospores* by air currents. *Puffballs* are globose basidiocarps that produce *basidiospores* completely within a protective covering and release them for wind dispersal by rupture of the cover. *Shelf fungi* are found growing on dead or damaged trees and logs. *Basidiospores* are produced within tubes on their lower surface.

Color the spermatium (I), receptive hypha (J), and teliospore (K), as well as previously listed structures in the illustrations of Heterobasidiomycetidae sexual reproduction at the bottom of the plate.

Most Heterobasidiomycetidae produce individual haploid cells called *spermatia* that fuse with a *receptive haploid hypha* of the opposite mating strain to form a dikaryotic cell that then produces *dikaryotic hyphae.* None produce basidiocarps. Under suitable conditions, masses of *dikaryotic hyphae* produce thick-walled *teliospores* consisting of two binucleate cells each containing one nucleus of each mating strain. Fusion of the two nuclei within each cell forms two diploid cells. Upon germination of the *teliospore,* each cell produces a *basidium* within which meiosis of the diploid nucleus occurs to form four haploid nuclei within each *basidium.* Sterigmata form along the length of the *basidium,* and a single haploid nucleus and some cytoplasm migrate into each and form a *basidiospore,* which is released when mature.

CLUB FUNGI INTRODUCTION.

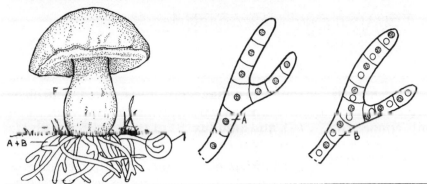

MYCELIUM A+B
HYPHA ★
  HAPLOID A
  DIKARYOTIC B

ASEXUAL ★
SPORE C

FRAGMENTATION

SPORULATION

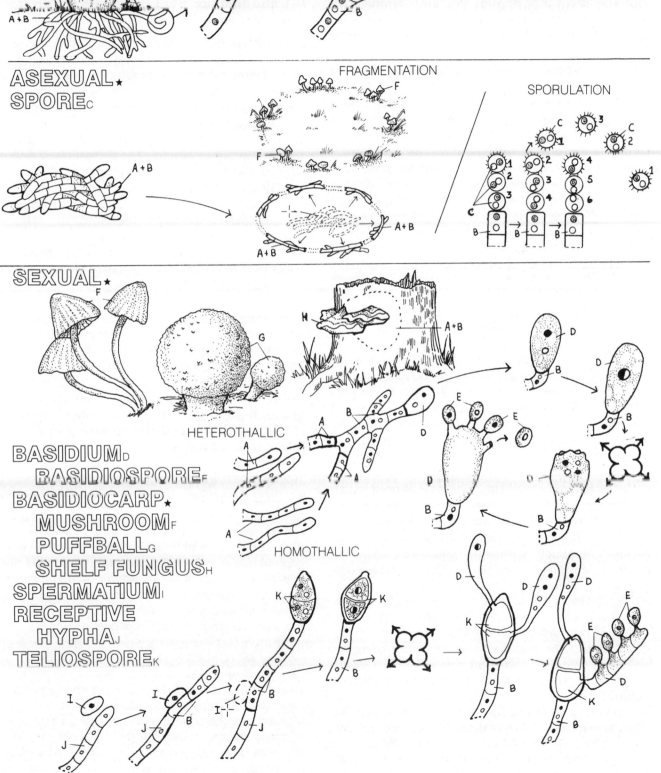

SEXUAL ★

HETEROTHALLIC

HOMOTHALLIC

BASIDIUM D
  BASIDIOSPORE F
BASIDIOCARP ★
  MUSHROOM F
  PUFFBALL G
  SHELF FUNGUS H
SPERMATIUM I
RECEPTIVE
  HYPHA J
TELIOSPORE K

34
FLESHY FUNGI (BASIDIOMYCETES)

Color the basidiospores (A), haploid hyphae (B), and dikaryotic hyphae (C) at the top of the plate and the diagrams depicting clamp connection formation at the right. Note that there are haploid and dikaryotic hyphae in the mycelium at top left.

A new generation of mushrooms begins with the formation of meiospores, called *basidiospores*, in club fungi. *Basidiospores* of two different mating strains are produced. *Basidiospores* depend primarily upon wind for dispersal, and if a *basidiospore* lands in a favorable location, it may germinate and produce a branching haploid filament, the *haploid hyphae,* that will spread throughout a favorable substrate, forming an extensive haploid mycelium. Growth is through mitotic division of the apical cell of each hyphal branch. When a *haploid hypha* of one mating strain contacts a *haploid hypha* of another mating strain, the cells at the point of contact fuse. The two nuclei within a fused cell, one nucleus from each mating type, do not fuse. Instead, a binucleate, or dikaryotic, cell is formed. It divides to produce extensive and pervasive *dikaryotic hyphae,* or dikaryotic mycelium, within the substrate.

Because of the binucleate condition of the *dikaryotic hypha,* two nuclei must divide synchronously (at the same time) at each mitotic division to produce new dikaryotic cells. At each division, the apical cell assumes a specialized configuration, called the clamp connection (figures 1–4), which is formed as an apparent adaptation to ensure sufficient space for binuclear division.

Development of the clamp connection begins with the formation of a rearward-directed pouch on the apical cell (figure 1). The two nuclei of the dikaryotic apical cell become positioned so that one nucleus divides completely within the apical cell while the other nucleus divides into the pouch at an angle to the axis of the filament. A new cell wall forms in the *dikaryotic hypha* tip, leaving one new nucleus in the cell adjacent to the new apical cell (figure 2), which, including the nucleus in the pouch at the top, now has three nuclei. The pouch of the apical cell elongates and bends toward the adjacent uninucleate cell (figure 2). When the elongating pouch contacts the adjacent uninucleate cell, an opening is formed in the walls between the two cells at the point of contact and the nucleus in the pouch is transferred to the adjacent uninucleate cell (figure 3). As the last step in the division process (figure 4), a wall forms between the pouch and the new apical cell. In this manner, the binucleate integrity of the *dikaryotic hypha* is maintained. All the dikaryotic cells are formed in this manner.

Color the dikaryotic hyphae (C), button stage mushrooms (D) and the stipe (E), pileus (F), gills (G), and annulus (H) of the mushroom cluster illustrated.

After the *dikaryotic hyphae* have formed an extensive mycelium to serve as an energy base and if environmental conditions are favorable, the mycelium begins to form a fruiting body, or mushroom, that grows above the surface of the substrate. The mushroom begins as a small mass, or *button,* of densely compacted hyphae. These hyphae quickly proliferate and the mushroom grows rapidly. The mushroom is only the "tip of the iceberg"; it is the sexual reproductive structure of a much more extensive subsurface vegetative plant that consists of an extensive mycelium. As the mushroom develops, the *stipe,* or stalk, and *pileus,* or cap, begin to take shape.

While the mushroom is immature and pushing its way upward through the substrate, the spore-forming *gills* are protected by a covering, the inner veil (not shown), which connects the *pileus* and *stipe.* As the mushroom approaches maturity, the inner veil pulls free from the *pileus* and remains on the *stipe* as a ring of tissue called the *annulus.* The expansion of the *pileus* exposes the *gills.*

Color the large single mushroom and the diagrams depicting gill (G) sections showing basidia (I) and the basidium maturation sequence on the left margin of the page.

The surface of the *gills* is covered with a multitude of club-shaped cells called *basidia.* Initially, the *basidia* are dikaryotic. As the mushroom matures, the two nuclei within each *basidium* undergo karyogamy, or nuclear fusion, to form a diploid nucleus. Meiosis of the diploid nucleus within each *basidium* produces four haploid nuclei, two of each mating strain. These nuclei migrate to the tip of the *basidium,* which forms four pouches. Each haploid nucleus enters a pouch and forms a *basidiospore,* which is situated at the tip of a stalklike sterigmatum (not separately colored). As the mushroom matures and the *pileus* expands, the *basidiospores* are released to start a new generation. Most of the *basidiospores* drop directly below the *pileus,* but some are carried aloft by the wind and deposited at some distant location, providing a means of dispersal.

Some asexual reproduction occurs in mushrooms through fragmentation of the mycelial hyphae of either the haploid or dikaryotic stage. Separation of a mycelium, called spawn, is used as a means of propagating mushrooms in commerce.

MUSHROOM
LIFE
HISTORY.
BASIDIOSPORE A
HAPLOID HYPHA B
DIKARYOTIC HYPHA C

MUSHROOM ★
BUTTON STAGE D
STIPE E
PILEUS F
GILL G
ANNULUS H
BASIDIUM I

CLAMP CONNECTION

BASIDIOSPORE
FORMATION

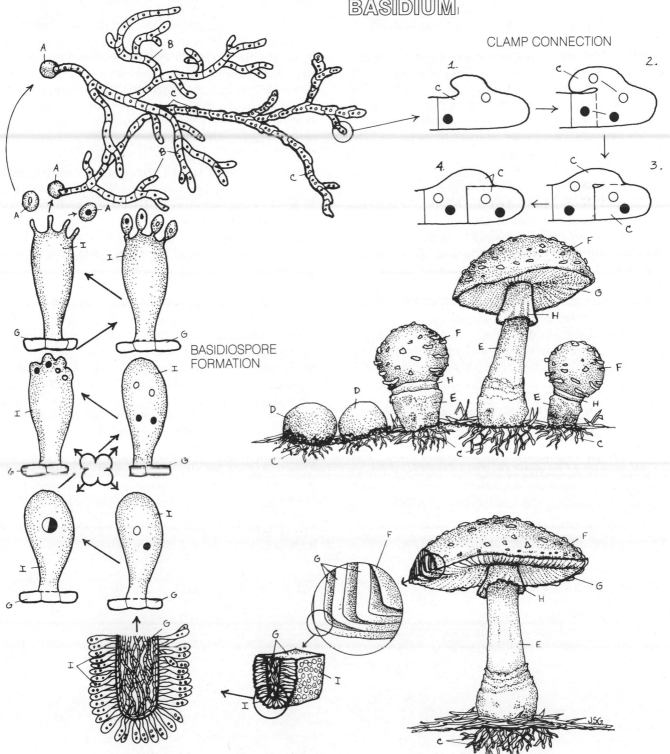

35

WHEAT RUST (BASIDIOMYCETES) BARBERRY HOST

The highly destructive wheat rust fungus, *Puccinia graminis* variety *tritici*, provides one example of the complex life cycles typical of many parasitic Heterobasidiomycetidae. In wheat rust, two different hosts, wheat (*Triticum*) and barberry (*Berberis*), are required for life cycle completion. Wheat rust cannot survive without both hosts present. This two-plate survey begins with the infection of a barberry plant by *basidiospores* in the spring.

Color the upper epidermis (A), mesophyll (B), and lower epidermis (C) of the barberry leaf, using three shades of green. Also color the basidiospores (D), haploid hyphae (E), spermagonia (F), spermatophores (G) and spermatia (G¹), and the receptive hyphae (H). Stop when you finish with the largest diagram, depicting spermatia (G¹) release, though a portion of it will be uncolored.

Infection of the barberry plant begins with the arrival of *basidiospores* on barberry leaf surfaces. These germinate and send *haploid hyphae* through the *upper* and *lower epidermis* layers into the *mesophyll*, or inner, leaf tissues. The *haploid hyphae* proliferate, using nutrients taken from the leaf tissues and soon form an extensive haploid mycelium within the leaf. When a sufficient amount of haploid mycelium has formed, many flask-shaped, blisterlike pustules, the *spermagonia* (pycnia), develop on the upper surface of the leaf. These eventually break through the *upper epidermis* and open as they mature. *Spermagonia* are organized masses of *haploid hyphae* that contain two types of specialized sexual reproductive structures. *Spermatophores* are terminal *haploid hyphae* within

a *spermagonium* that continually produce small, individual cells, called *spermatia* (pycniospores), from their tips by mitosis. Each *spermagonium* releases large numbers of *spermatia*, which are then dispersed by wind or insects. *Receptive hyphae*, the second type of sexual reproductive structure, are long *haploid hyphae* attached within a *spermagonium* and exerted through its opening at maturity.

Color the dikaryotic hyphae (I), aecium (J), aeciospore-forming hyphae (K), and aeciospores (L), as well as the previously listed structures in the remaining diagrams.

Wheat rust fungus has two mating strains, designated plus and minus, that are each self-incompatible. Therefore, *spermatia* and *receptive hyphae* produced by the same mating strain will not fuse. When a dispersed *spermatium* of one mating strain contacts a *receptive hypha* of the opposite mating strain, fusion occurs and a dikaryotic cell is formed. The *dikaryotic hypha* that develops reinvades the *mesophyll* tissues and forms a dikaryotic mycelium within the leaf intermingled with the haploid mycelium. When a sufficient amount of dikaryotic mycelium has formed, many globose, blisterlike pustules, the *aecia*, develop on the lower surface of the barberry leaf. As they mature, they enlarge, push through the *lower epidermis*, and break open. *Aecia* are organized masses of *dikaryotic hyphae* that contain *aeciospore-forming hyphae*. These continually produce enormous numbers of *aeciospores* from their tips by mitosis. The binucleate, or dikaryotic, *aeciospores* are dispersed by wind to the second host, wheat.

WHEAT RUST LIFE HISTORY.

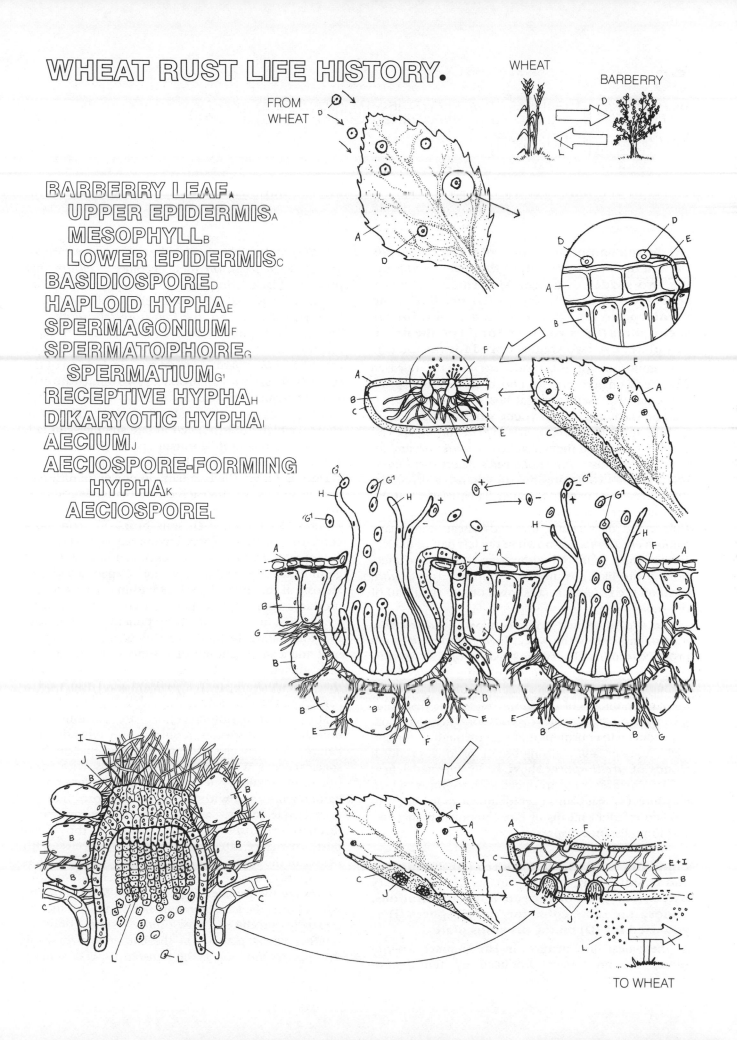

WHEAT

BARBERRY

FROM WHEAT

BARBERRY LEAF A
UPPER EPIDERMIS A
MESOPHYLL B
LOWER EPIDERMIS C
BASIDIOSPORE D
HAPLOID HYPHA E
SPERMAGONIUM F
SPERMATOPHORE G
SPERMATIUM G1
RECEPTIVE HYPHA H
DIKARYOTIC HYPHA I
AECIUM J
AECIOSPORE-FORMING HYPHA K
AECIOSPORE L

TO WHEAT

36
WHEAT RUST (BASIDIOMYCETES) WHEAT HOST

On the top half of the plate, color the epidermis (A) and internal tissue (B) of the wheat stalk, using two shades of green. Also color the aeciospores (C), dikaryotic hyphae (D), uredia (E), and urediospores (F). Choose the same color for the aeciospores (C) as you used for (L) on the previous plate. Use two shades of red for (E) and (F).

The multitude of dikaryotic *aeciospores* released from aecia on barberry plant hosts during early spring are wind dispersed to wheat hosts. *Aeciospores* that land on the *epidermis* of young wheat plants quickly germinate and send *dikaryotic hyphae* through pores, called stomates, in the wheat host's *epidermis* and into the wheat host's *internal tissues*. There the *dikaryotic hyphae* proliferate into an extensive mycelium that extracts nutrients through haustorial (food extracting) connections into the wheat host's cells. Uninfected or recently infected wheat stems have smooth *epidermis* (as shown in the left half of the late spring wheat stalk) but within two weeks after infection, blisterlike pustules or eruptions, the *uredia*, form on the *epidermis* (as shown on the right half of the stalk).

The *uredia*, which begin as masses of *dikaryotic hyphae* just below the wheat host's *epidermis*, soon produce terminal dikaryotic *urediospores* by mitosis. When near maturity, the developing *uredia* burst through the wheat host's *epidermis* and begin releasing large numbers of *urediospores*, whose rust color give wheat rust its name. *Urediospores* are continually produced in enormous numbers through spring and early summer, and a strong breeze produces reddish clouds of *urediospores* above badly infected wheat fields. The asexually produced *urediospores*, which may infect wheat plants in epidemic proportions, provide an efficient means of rapid population increase and long-distance dispersal.

Color the telia (G), teliospores (H), basidia (I), and basidiospores (J), as well as the previously listed structures, on the remaining illustrations. Choose the same color for basidiospores (J) as you used for (D) on the previous plate.

As the wheat host matures in late summer, *urediospores* are no longer produced by the *uredia*.

Urediospores cannot survive the cold temperatures of winter. Each *uredium* then becomes a *telium* that produces black *teliospores*. In teliospore formation, elongated, terminal dikaryotic cells within the *telium* form an enlarged end, the immature *teliospore* containing two haploid nuclei. A septum forms to separate this immature *teliospore* from the stalk portion of the *dikaryotic hypha*. A single mitotic division of the terminal dikaryotic cell produces two binucleate cells at each hyphal tip that form thick, resistant black walls to become a two-celled *teliospore*. The *teliospores*, which are produced in late summer and fall, function as resistant, dormant spores that carry the wheat rust fungus through winter.

During winter, the two haploid nuclei, of different mating strains, within each of the two cells comprising the *teliospore* fuse to form two diploid cells, or zygotes. Therefore, each stalk bears one *teliospore* consisting of two zygotes surrounded by a thick, resistant wall. In early spring, each cell of a *teliospore* germinates and produces a club-shaped *basidium*. Meiosis of the diploid nucleus within each *basidium* produces four haploid nuclei, two of each mating strain, that migrate into four pouchlike sterigmata that form along the length of each *basidium*. A spore wall and a small amount of cytoplasm forms around each nucleus to produce four haploid *basidiospores* that, when mature, are released and wind dispersed to barberry plants in early spring.

Most Heterobasidiomycetidae are parasitic, and both autoecious species (requiring only one host) and heteroecious species (requiring two separate host species) occur. Wheat rust is heteroecious, with a portion of the sexual cycle, syngamy, taking place on the barberry host and with the remaining portion of the sexual cycle, meiosis, as well as asexual reproduction, through *urediospores*, taking place on the wheat host. To ensure infection of both host species, both phases of the wheat rust fungus must produce numerous spores capable of efficient dispersal. *Basidiospores* are the transfer dispersal units formed on the wheat hosts; the *teliospores* are not dispersed, and the asexually produced *urediospores* serve only to reinfect wheat plants. On the barberry plants, the *aeciospores* function as the transfer dispersal units.

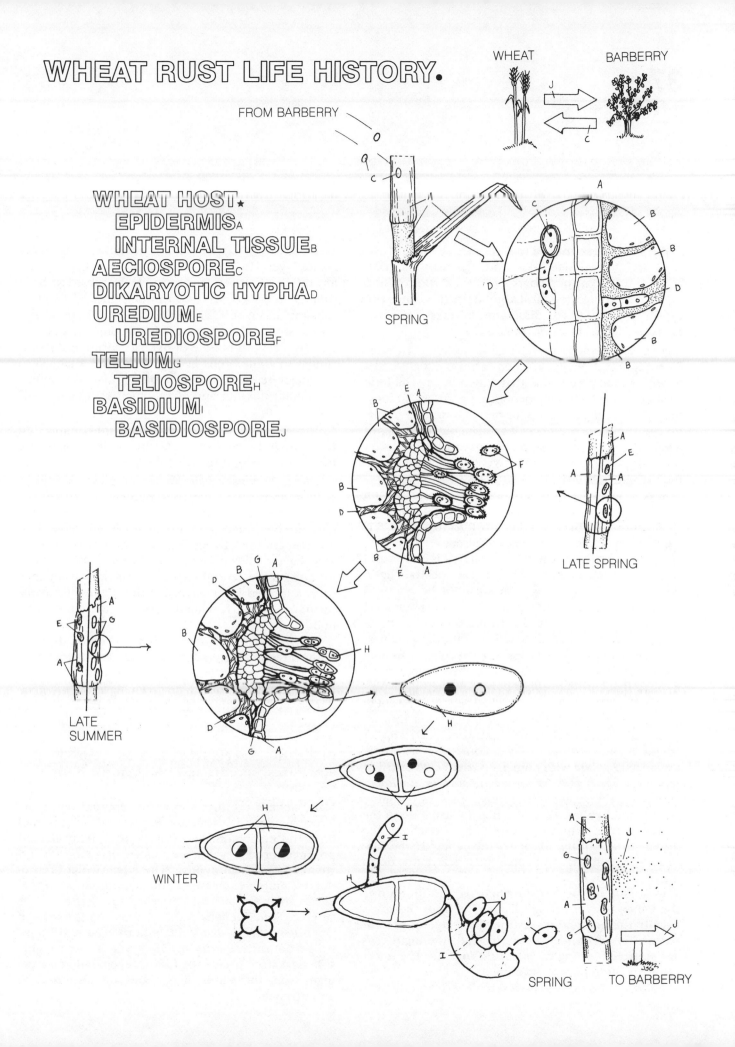

WHEAT RUST LIFE HISTORY.

WHEAT HOST★
EPIDERMISA
INTERNAL TISSUEB
AECIOSPOREC
DIKARYOTIC HYPHAD
UREDIUME
UREDIOSPOREF
TELIUMG
TELIOSPOREH
BASIDIUMI
BASIDIOSPOREJ

Color all diagrams labeled "crustose" (A), "foliose" (B), "fruticose" (C), and "gelatinous" (D), as well as the upper cortex (E), medulla hyphae (F), lower cortex (G), and algal cells (J) within the medulla on the two diagrams labeled "heteromerous" and "homoiomerous."

Lichens are formed through a combination of two different kinds of plants, a lichen-forming fungus and a green or blue-green alga, that function in symbiotic cooperation as mutually benefiting partners. In most lichens, the fungal species, or mycobiont, cannot survive without the presence of an alga; but a number of algal species, or phycobionts, seem capable of surviving without a fungus. The mycobiont provides the bulk of the lichen tissue in all cases.

Several growth forms of lichens, based primarily on external morphology, are recognized; but the *crustose, foliose, fruticose,* and *gelatinous* forms are perhaps the most commonly encountered.

Crustose lichens form a thin crust that is closely and tightly appressed and adherent to the substrate upon which they are growing. They cannot be removed from the substrate in one piece. Two morphological types of *crustose* lichens, heteromerous and homoiomerous, are recognized, based on internal organization. In heteromerous types, the mycobiont usually consists of highly gelatinous, tightly compressed hyphae that form an *upper cortex* in the upper portion of the *crustose* lichen. No algal cells are present in the *upper cortex,* which may be colored or translucent. Immediately below the tightly compacted *upper cortex* is the *medulla,* a layer consisting of loosely organized *hyphae* with numerous *algal cells* distributed in the upper part. The lowermost layer, the *lower cortex,* is a thin, compacted layer, consisting of tightly compacted hyphae, that is similar in structure to the *upper cortex.* Rhizoids scattered throughout the lower surface of the *lower cortex* penetrate the substrate for anchorage but apparently derive few or no nutrients from it.

In homoiomerous *crustose* lichens, stratification of the lichen body is less apparent. The *upper cortex* is usually thinner, and no distinct algal layer is present within the *medulla hyphae*. Instead, the *algal cells* are scattered among the *medulla hyphae*. The *lower cortex* is often poorly developed.

Foliose lichens closely resemble *crustose* lichens and consist of *upper cortex,* medulla, and *lower cortex* layers. However, the *foliose* lichen body is usually thicker, is raised farther above the substrate, has a thinner *lower cortex,* and can be removed from its substrate more or less intact because it adheres less to the substrate. Many *foliose* lichen bodies are attached only by a central bundle of hyphae, forming rhizoids similar to an umbrella handle; others are attached to the substrate by scattered bundles of hyphal rhizoids. Both heteromerous and homoiomerous forms exist.

Color the two halves of the diagram illustrating the internal structure of fruticose lichens in the bottom right quadrant of the plate.

Fruticose lichens are simply to complexly branched with cylindrical to flattened branches that may be erect to pendulous and more than twenty feet long. Most are heteromerous with hollow branches or with a compact core of fungal tissue, the *central cortex.* Most have a thick *outer cortex.* In most, the *medulla hyphae* fill the central core of a branch, or they may form a limited layer enclosing the *algal cells* inside the *outer cortex,* leaving the central core completely hollow.

Gelatinous lichens are more or less homoiomerous, with abundant blue-green algae and very little fungus present. Most are black or bluish black, similar in appearance to some blue-green algae, highly gelatinous, and sometimes appearing as tarlike blotches on rocks.

Color the specialized asexual reproductive structures, soredia and isidia, at the bottom of the plate.

Many lichens produce specialized asexual reproductive structures that function in the co-distribution of both the mycobiont and phycobiont. Many lichens produce soredia, which are clusters of *algal cells* surrounded by fungal hyphae. These lightweight reproductive units are released through lesions in the *upper cortex* that permit air currents to distribute the soredia. Some lichens produce small outgrowths of the *upper cortex* that contain both algal and fungal cells from the *medulla hyphae* layer. These roughly spherical structures, called isidia, eventually disarticulate from the parent lichen and are dispersed by air currents.

LICHENS.

MYCOBIONT ★
UPPER CORTEX E
MEDULLA HYPHAE F
LOWER CORTEX G
OUTER CORTEX H
CENTRAL CORTEX I
PHYCOBIONT ★
ALGAL CELLS J

CRUSTOSE A

FOLIOSE B

FRUTICOSE C

GELATINOUS D

HETEROMEROUS

HOMOIOMEROUS

SOREDIA

ISIDIA

38
ALGAL HABITATS

Using a shade of green, color the representatives of the green algae (A) and their general area of distribution.

Green algae, the most widely distributed of all algal divisions, are found from the snowy peaks of mountain tops to a maximum depth of about a hundred meters in warm oceans. Many species are terrestrial (land dwelling), but they are most abundant in freshwater aquatic habitats. Most terrestrial species occur in continuously or frequently moist soil.

In aquatic habitats, free-floating (suspended) motile and nonmotile *green algae* are a minor component of freshwater plankton (microscopic floating organisms) and are practically absent in marine plankton. Species characteristically attached to surfaces, sessile *green algae,* are the most abundant forms. Freshwater habitats, such as streams, rivers, ponds, and lakes, contain the greatest number (about 90 percent) of *green algae* species. Most green "pond scum" is usually a floating mass of filamentous (hairlike) *green algae* of several species. In aquatic habitats with warm water and a high nitrogen content coupled with high light levels, as during the summer months, a tremendous population increase, or "algal bloom," that clouds the water may occur. Algal blooms, which are primarily due to asexual reproduction, are most prevalent in polluted waters. Some other algal groups, especially dinoflagellates, may also produce blooms.

There are far fewer marine *green algae* species than freshwater species, and most occur as sessile forms in the intertidal zone (the area between high tide and low tide) because their photosynthetic pigments are most effective in red and blue light wavelengths, which do not penetrate into deep water.

Using a light shade of brown, color the representatives of the golden-brown algae (B) and their general area of distribution.

Golden-brown algae contain the green photosynthetic pigment chlorophyll, as do all groups of algae, but its color is masked by golden-brown accessory photosynthetic pigments. The pigments found in the *golden-brown algae* also utilize light wavelengths that do not penetrae into deep water; therefore, most species occur in the upper levels of aquatic habitat. Many *golden-brown algae* are planktonic, but sessile forms are also found. Plankton consists of small macroscopic and microscopic plant and animal life suspended in the upper levels of aquatic habitats, both freshwater and marine. The phytoplankton (plant) component of plankton consists primarily of *golden-brown algae* and dinoflagellates (not shown).

The *golden-brown algae,* as treated here, includes the diatoms. Diatoms are found in both freshwater and marine habitats in tremendous numbers. In diatoms, colonial forms as well as motile, unicellular forms are abundant. Light brown patches of "pond scum" often consist of masses of these *golden-brown algae.*

Using a shade of brown, color the representatives of the brown algae (C) and their general area of distribution.

With very few exceptions, *brown algae* are restricted to marine habitats. They are most abundant in temperate region oceans, but they are also well represented in tropical oceans. Because their photosynthetic pigments utilize light wavelengths that do not penetrate deeply in water, they are most abundant in shallow ocean waters, but some are found down to about a hundred meters. Most *brown algae* are found in the intertidal zone. *Brown algae* typically grow attached to a substrate, and those that are anchored in deeper water usually have gas-filled bladders that maintain the photosynthetic surface in the upper levels of the ocean for optimal photosynthetic efficiency.

Using a shade of red, color the representatives of the red algae (D) and their general area of distribution.

Like the *brown algae,* the *red algae* are primarily found in marine habitats. *Red algae* are most abundant in warm tropical oceans, but they are also present in cool temperate oceans. They range from the intertidal zone down to a maximum depth of about two hundred meters, but they are best represented in the subtidal zone below the low-tide level. The *red algae* occur at greater depths than any other algae due to the presence of accessory photosynthetic pigments that utilize light wavelengths that penetrate to those depths. *Red algae* are typically found attached to a substrate. Most are attached to inorganic substrates such as rocks, but many are epiphytic (found growing upon another plant) on other algae.

DISTRIBUTION OF MAJOR ALGAL GROUPS.

GREEN ALGAE_A
GOLDEN-BROWN ALGAE_B
BROWN ALGAE_C
RED ALGAE_D

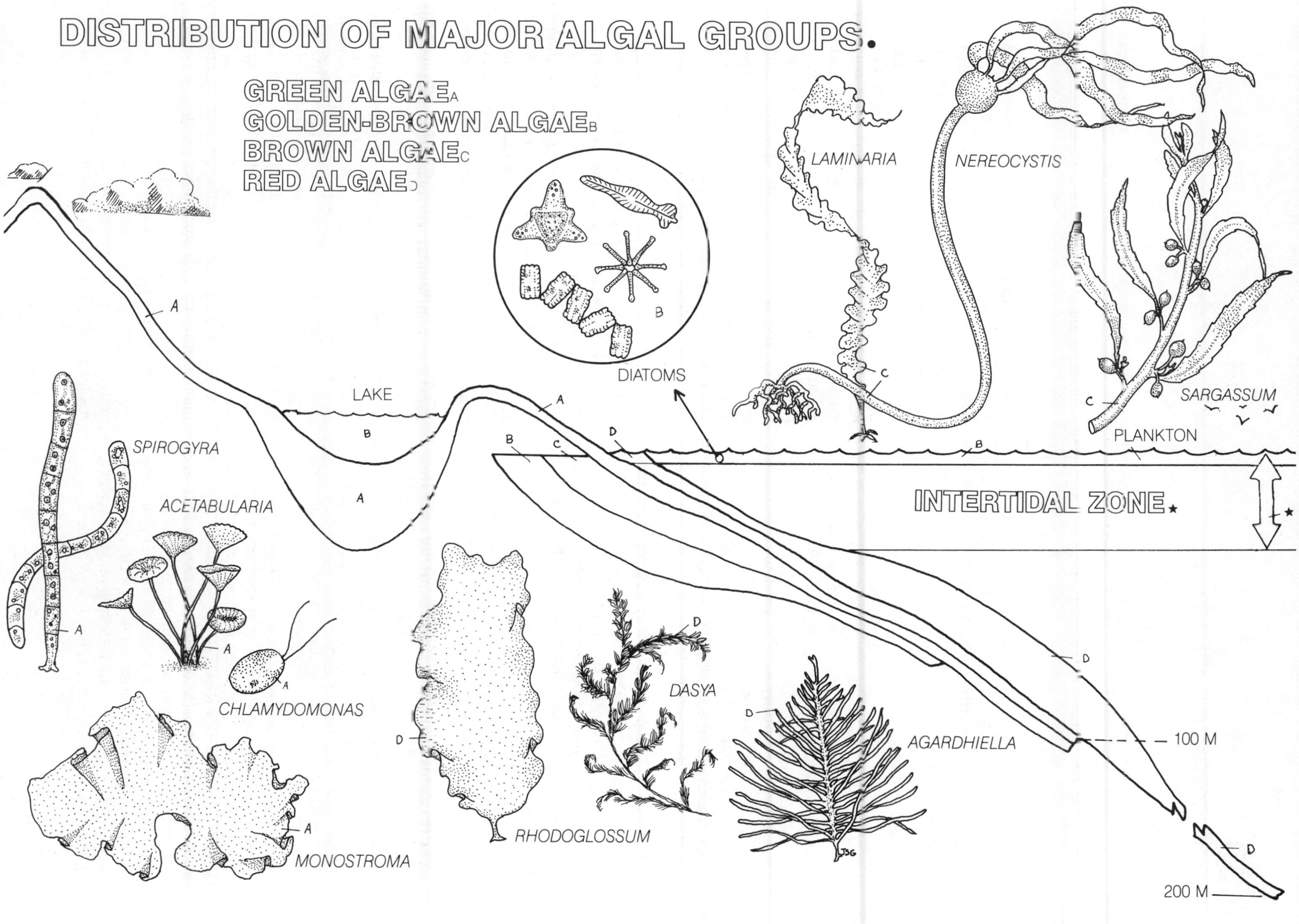

NEREOCYSTIS

LAMINARIA

SARGASSUM

PLANKTON

INTERTIDAL ZONE ★

100 M

200 M

AGARDHIELLA

DASYA

RHODOGLOSSUM

DIATOMS

LAKE

SPIROGYRA

ACETABULARIA

CHLAMYDOMONAS

MONOSTROMA

39
GREEN ALGAE (CHLOROPHYTA)

Color over the lines indicating individual flagella (A) on the diagrams illustrating flagellar arrangements of motile cells. In addition, color the unlabeled cell bodies a light shade of green.

Green algae, division Chlorophyta, are a large and widely distributed group exhibiting a broad range of morphological diversity. In all species known to undergo sexual reproduction, at least one of the two gametes entering syngamy is motile. Many species also produce motile cells during asexual reproduction, and some, such as *Chlamydomonas*, have motile vegetative cells. Motility in most green algae is due to apically attached whiplash *flagella*. That is, the *flagella* are inserted into the cell body on the anterior end and are smooth and whiplike. The rhythmic beating action of the *flagella* functions to pull the cells through the water. Cells with *two flagella* have the commonest flagellar arrangement in both reproductive and vegetative motile cells. Some species produce cells having *four flagella*; others produce cells having *eight flagella*; and a few produce cells with a Friar Tuck-like *ring* of small *flagella* around their anterior end.

Color the isogamous (B), anisogamous (C), and oogamous (D) gamete arrangements that occur in sexual reproduction in green algae. Color over the lines indicating flagella, using the same color chosen for the flagella (A) colored above.

Most green algal species, such as *Chlamydomonas* and *Ulva*, are *isogamous*; but some, such as *Bryopsis*, are *anisogamous*, and a few, such as *Spirogyra* and *Oedogonium*, are *oogamous*. Species exhibiting *isogamy* or *anisogamy* freely release gametes into the water. Large numbers of gametes of both sexes must be produced to ensure effective sexual reproduction. In *isogamy* and *anisogamy*, both gamete types are usually much smaller than a single vegetative cell due to the series of mitotic divisions by which they are formed, but in *oogamy*, a single large female gamete approaching the size of a vegetative cell is often produced. Therefore, in oogamous species, a greater amount of cytoplasm, with its ample nutrient reserve, is available to the zygote.

Color the diagrams illustrating isomorphic (E) and heteromorphic (F) alternations of generations.

Most green algae have a haplontic life history, but a few are diplontic or diplohaplontic. Haplontic and diplontic species do not exhibit an alternation of generations, as only a single generation, either haploid or diploid, comprises the vegetative stage. Two different types of alternations of generations occur in the few diplohaplontic green algae. In species having an *isomorphic* alternation of generations, such as *Ulva*, the vegetative forms of both the gametophytic and sporophytic generations are morphologically similar. On the other hand, the gametophytic and sporophytic generations are morphologically different in species such as *Derbesia*, which have *heteromorphic* alternation of generations.

Color the unicellular green alga (G) and the diagrams of the volvocine (H), tetrasporine (I), and siphonous (J) lines.

Three distinct lines, or evolutionary trends, thought to be derived from a *unicellular* chlamydomonaslike ancestor and primarily based on vegetative morphology, are evident in the green alage. Each line contains members that demonstrate a sequence of increasing complexity from unicells or few-celled filaments to complex multicellular or branched multinucleate forms. The *volvocine* line consists of motile *unicellular* and colonial forms, which may contain from few to tens of thousands of cells.

The *tetrasporine* line consists of nonmotile, multicellular forms, often forming complex branched filaments composed of numerous uninucleate cells held together by a cementing layer between adjacent cell walls. Each mitotic division produces two uninucleate cells.

Members of the *siphonous* line consist of one or more multinucleate cells. If karyokinesis is not followed by cytokinesis, formation of multinucleate cell results, as in *Derbesia* and *Bryopsis*. If a progression of nuclear divisions within a multinucleate cell is occasionally accompanied by a cytoplasmic division, a series of multinucleate cells result, as in the alga *Cladophora*.

GENERAL MORPHOLOGY.

WHIPLASH FLAGELLA ★
TWO APICAL $_{A^1}$
FOUR APICAL $_{A^2}$
EIGHT APICAL $_{A^3}$
RING $_{A^4}$

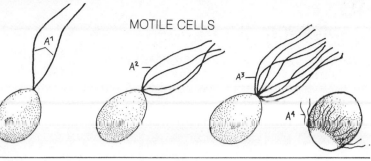

MOTILE CELLS

SEXUAL REPRODUCTION ★
ISOGAMY $_B$
ANISOGAMY $_C$
OOGAMY $_D$

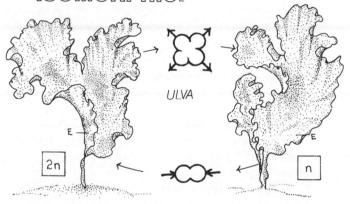

GAMETES

GENERATIONS ★
ISOMORPHIC $_E$

ULVA

2n   n

HETEROMORPHIC $_F$

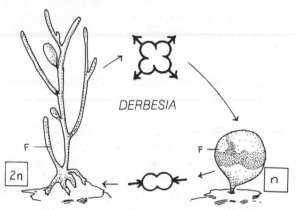

DERBESIA

2n   n

EVOLUTIONARY LINES ★
UNICELLULAR $_G$
VOLVOCINE $_H$
TETRASPORINE $_I$
SIPHONOUS $_J$

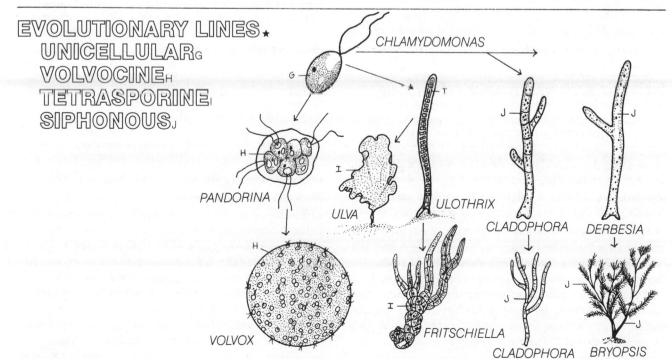

CHLAMYDOMONAS

PANDORINA

ULVA

ULOTHRIX

CLADOPHORA

DERBESIA

VOLVOX

FRITSCHIELLA

CLADOPHORA

BRYOPSIS

40
GREEN ALGAE (CHLOROPHYTA)

Color the cell wall (A), cytoplasm (B), chloroplast (C), pyrenoid (D), starch grains (E), eye spot (F), nucleus (G), nucleolus (H), mitochondria (I), contractile vacuole (J), and flagella (K) on the diagram of Chlamydomonas at the top of the plate. Long flagella are shortened by breaks in the drawing.

Chlamydomonas is a minute, unicellular freshwater green alga with a cellulosic *cell wall* surrounding the plasma membrane-(not shown) bound protoplast. The *cytoplasm* contains a single, cup-shaped *chloroplast.* Within the *chloroplast* is at least one *pyrenoid*, which are small structures found within some algal *chloroplasts* that apparently function in the formation of *starch grains*, and a single red *eye spot*, a small organelle that functions in phototaxic responses, or light-induced movements, by the cell. Depending upon light quality and quantity, *Chlamydomonas*, and most other motile green algae, move either toward or away from a light source. The *nucleus*, and the contained *nucleolus*, typically reside within the hollow formed by the cup-shaped *chloroplast. Mitochondria* are found scattered in the *cytoplasm* but concentrated around the flagella bases. One or two *contractile vacuoles*, which function in the removal of excess water, are usually found toward the anterior, or flagellar, end. Two equal whiplash *flagella*, which function to pull the cell through the water by their beating motion, are inserted at the anterior end of the cell.

Color the two mature vegetative cells (L), which are now sexually receptive as functional gametes (M), located in the twelve o'clock position on the circular diagram depicting the sexual reproductive cycle. Continue coloring until you finish with the first diagram of a zygote (N).

Chlamydomonas exhibits a haplontic, or zygotic, life history in which the haploid *vegetative cells* and sex cells, or *gametes*, are morphologically identical. In fact, under specific environmental conditions, haploid *vegetative cells* simply alter their behavior and function as *gametes*. The life history is also hetcrothallic since two opposite mating strains, called plus (+) and minus (−) because both sexes are morphologically identical, are required for sexual union, or syngamy, to occur. In sexual reproduction, two *gametes* of opposite mating strains approach one another by their anterior ends, entwine their *flagella*, and pull close together. At this time, a thin strand of interconnecting cytoplasm (not shown) often forms between the two cells. Soon the two *gametes* merge, retract their *flagella*, and form a single, nonmotile *zygote*.

Continue coloring until you finish with the diagram showing the half-grown vegetative cells (L).

Soon after *zygote* formation, the *zygote* produces a thick cellulosic wall to become a zygospore, a dormant, resistant spore stage that has a broad range of tolerance for many environmental parameters, especially extremes of moisture and temperature. The zygospore carries *Chlamydomonas* through its stay in the dry, caked mud at the bottom of a dried-up pond. In this stage, dispersal may occur through mud carried on the feet of aquatic birds and other animals. When environmental conditions favorable for the growth of haploid *vegetative cells* return, the diploid nucleus within the zygospore undergoes meiosis to produce four haploid nuclei (not shown). Following meiosis, one or more synchronous mitotic divisions produce numerous haploid nuclei. A cell wall then forms about each haploid nucleus, along with a small amount of cytoplasm, to form haploid *sexual zoospores*. Rupture of the zygospore wall releases the enclosed *sexual zoospores,* or miniature haploid *vegetative cells,* which then grow to mature size. Population increase is effected through meiosis and mitosis, and the motility of the *sexual zoospores* accounts for some dispersal.

Finish coloring the plate. This portion illustrates asexual reproduction.

After the *sexual zoospores* undergo growth, through cell enlargement, to become mature cells, they often form *mitosporangia* and lose their *flagella*. The protoplast within the *mitosporangium* undergoes two or more mitotic divisions to produce numerous *asexual zoospores*, or mitospores. When released from the *mitosporangium* through rupture of the thin mitosporangial wall, which is the old vegetative parent cell wall, the enclosed *asexual zoospores* function in population increase and dispersal. Continued asexual reproduction produces a rapid population increase.

CHLAMYDOMONAS LIFE HISTORY.

CHLAMYDOMONAS CELL★
CELL WALL A
PROTOPLAST★
  CYTOPLASM B
  CHLOROPLAST C
    PYRENOID D
    STARCH GRAIN E
    EYE SPOT F
NUCLEUS G
  NUCLEOLUS H
MITOCHONDRION I
CONTRACTILE VACUOLE J
FLAGELLUM K

ASEXUAL
  REPRODUCTION ★
MITOSPORANGIUM P
ASEXUAL ZOOSPORE Q

SEXUAL
  REPRODUCTION ★
VEGETATIVE CELL L
GAMETE M
ZYGOTE N
SEXUAL ZOOSPORE O

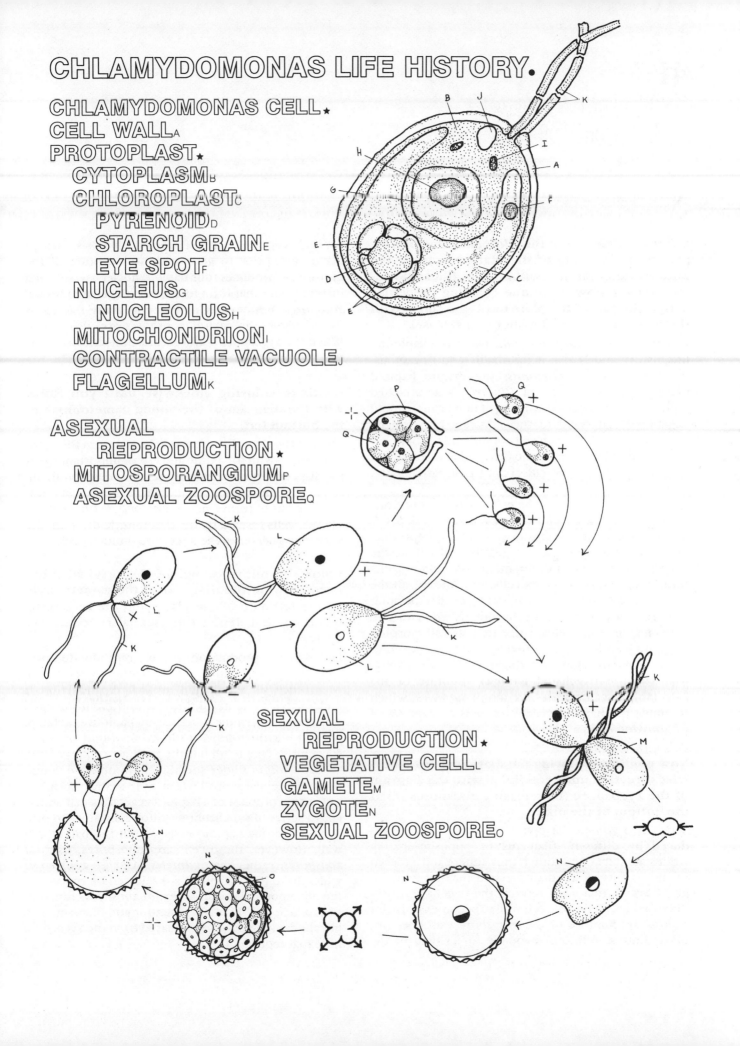

41

GREEN ALGAE (CHLOROPHYTA)

Color the diagram of the *Ulothrix* gametophyte (A) on the left side of the plate, but leave the mitosporangium (F) within the filament uncolored for now. Continue coloring clockwise across the top of the plate until you finish with the diagram depicting gamete (C) release.

Ulothrix, like *Chlamydomonas,* exhibits a haplontic life history and is also heterothallic and isogamous. But unlike *Chlamydomonas,* the mature haploid phase, or gametophyte, of *Ulothrix* is an attached multicellular filament consisting of numerous haploid vegetative cells held to the substrate by a single, sterile holdfast cell. Each cell comprising the filament is surrounded by a cellulosic cell wall and contains a single, bracelet-shaped chloroplast that partially and loosely surrounds the nucleus.

Each gametophyte produces *gametes* of only one mating strain, either plus (+) or minus (−), depending on its genetic composition. In sexual reproduction, one or more haploid vegetative cells along the length of the filament differentiate to become *gametangia,* or gamete-forming cells. In *Ulothrix,* all the haploid vegetative cells in a filament, except the holdfast cell, are capable of becoming *gametangia. Gametes,* or isogametes, since they are all morphologically identical, are formed by five or six synchronous mitotic divisions resulting in thirty-two or sixty-four small, biflagellated *gametes.* Rupture of the gametangial wall releases the motile *gametes,* which resemble the haploid vegetative cells or *gametes* of *Chlamydomonas.*

Now continue coloring the diagrams in a clockwise direction until you finish with the diagram of the zygospore (D) releasing zoospores (E) at the bottom of the plate.

When two *gametes* of opposite mating strains, produced by different filaments of opposite mating strains, contact one another, they adhere and undergo fusion to form a quadriflagellated (having four flagella) *zygote* that soon retracts the flagella and produces a thick, cellulosic cell wall to become a dormant *zygospore.* Rupture of the *zygospore* wall does not occur until conditions favorable for growth of the

haploid vegetative phase return. Meiosis, which occurs just prior to germination or rupture of the *zygospore,* produces four haploid nuclei, two of each mating strain, that form four quadriflagellated *sexual zoospores.* Besides having twice as many flagella as the *gametes,* the *sexual zoospores* are also larger. When the *zygospore* wall ruptures, the enclosed *sexual zoospores* are released.

Continue coloring clockwise until you finish with the diagram of the young gametophyte at the bottom left.

Though the *sexual zoospores* released from the *zygospore* may swim for a time and are an important agent for dispersal, they eventually settle and attach themselves to a suitable substrate, where they begin mitotic divisions to form a new *gametophyte* filament. Of the two cells formed by the first mitotic division, the lowermost becomes the anchoring holdfast cell.

Color the mitosporangium (F), originally left uncolored on the diagram of the gametophyte on the left side of the plate, and the remaining diagrams that depict the asexual reproductive cycle.

Like sexual reproduction, asexual reproduction provides a means of significant population increase. The occurrence of sexual and asexual reproduction is under environmental influence, and environmental conditions determine whether the haploid vegetative cells of a *gametophyte* form *gametangia* or *mitosporangia.* One to all but the holdfast cell may form *mitosporangia* under specific environmental conditions. In asexual reproduction, as in sexual reproduction, the protoplast of a haploid vegetative cell undergoes a series of synchronous mitotic divisions to produce numerous haploid cells within the parent cell wall. However, these cells are mitospores, *asexual zoospores,* rather than *gametes. Asexual zoospores,* which have four flagellae, are released by rupture of the mitosporangial wall. *Asexual zoospores* function to produce new haploid gametophytic filaments just like the *sexual zoospores* released from the *zygospore* in sexual reproduction.

ULOTHRIX LIFE HISTORY.

SEXUAL REPRODUCTION.
    GAMETOPHYTE_A
    GAMETANGIUM_B
    GAMETE_C
    ZYGOTE/ZYGOSPORE_D
    SEXUAL ZOOSPORE_E

ASEXUAL REPRODUCTION.
    MITOSPORANGIUM_F
    ASEXUAL ZOOSPORE_G

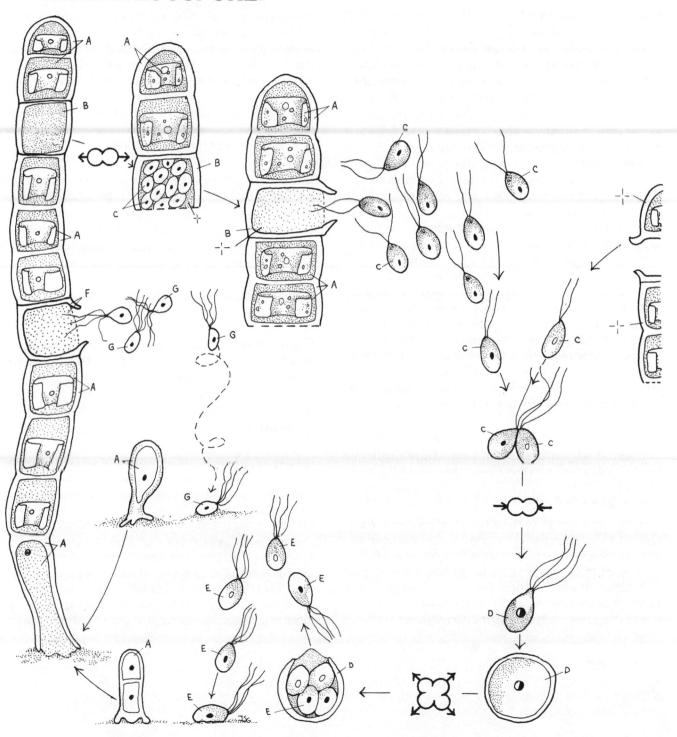

42
GREEN ALGAE (CHLOROPHYTA)

Color the sporophytic plant (A) in the upper left corner using a shade of green, and continue coloring clockwise across the top of the plate until you finish coloring the diagrams illustrating sexual zoospores (C). Leave the mitosporangium (H) and diploid zoospores (I) uncolored for now.

Ulva, or sea lettuce, is a common intertidal marine green alga. This lettuce-green, sheetlike alga is heterothallic, slightly anisogamous, and diplohaplontic with an isomorphic alternation of generations. The diploid generation, or *sporophyte,* is a delicate-appearing thin sheet, usually two cells thick and often more than a decimeter across, that is relatively tough and well adapted for survival in intertidal currents. It typically grows attached to a substrate by a strong holdfast.

In the sexual reproductive cycle, individual cells near the margin of the *sporophyte,* when induced by environmental conditions, develop into *meiosporangia* that produce meiospores, which mature into quadriflagellated *sexual zoospores.* Upon rupture of the outer wall of a meiosporangial cell, the enclosed *sexual zoospores,* which function as major dispersal units, are released. After swimming for a time, the *sexual zoospores* eventually settle on a substrate and become attached by their anterior end.

Color the young gametophytic plants (D), using a second shade of green, and continue until you finish coloring the diagrams illustrating gamete (F) release. Leave the unfertilized gamete (F¹) uncolored for now.

Germination and growth of the attached *sexual zoospores,* through repeated mitotic divisions, produces a haploid generation, or *gametophyte,* that is morphologically similar to the *sporophytic plant,* though of a different chromosome level. Under proper environmental conditions, individual cells along the margin of the *gametophytic plant* differentiate to become *gametangia.* A series of mitotic divisions of the gametangial protoplast produces numerous *gametes* within each gametangial cell. Biflagellated anisogametes, the female being slightly larger and greener than the male, are released by rupture of the gametangial wall. Since *Ulva* is heterothallic, each *gametophytic plant* produces either male or female *gametes,* not both. Because of this, *gametophytic plants* are often referred to as being either male or female.

Color the gametes (F) in the process of syngamy and continue coloring up the side of the plate until all components of the sexual cycle are colored.

Fusion occurs between male and female *gametes* from two different *gametophytes.* A quadriflagellated, motile *zygote* is initially formed, but it soon retracts its flagella to become a nonmotile *zygote.* Through repeated mitotic divisions, a new *sporophytic plant* is produced.

Color the mitosporangium (H) and the asexually produced zoospores (I) that function in asexual reproduction of the diploid generation. Also color the unfertilized gametes (F¹) that may function in asexual reproduction of the haploid generation.

Asexual reproduction has been observed for some species of *Ulva.* In the *sporophyte* of some species, marginal cells may become *mitosporangia* that produce quadriflagellated diploid *asexual zoospores* asexually by mitosis. These diploid *asexual zoospores* function much like the sexually produced haploid *sexual zoospores* except that they develop into new *sporophytes* rather than *gametophytes.* *Gametes* produced by the *gametophyte* of some species have been observed to function as haploid *zoospores.* These become attached to a substrate and undergo repeated mitotic divisions to form new *gametophytic plants.*

ULVA LIFE HISTORY.

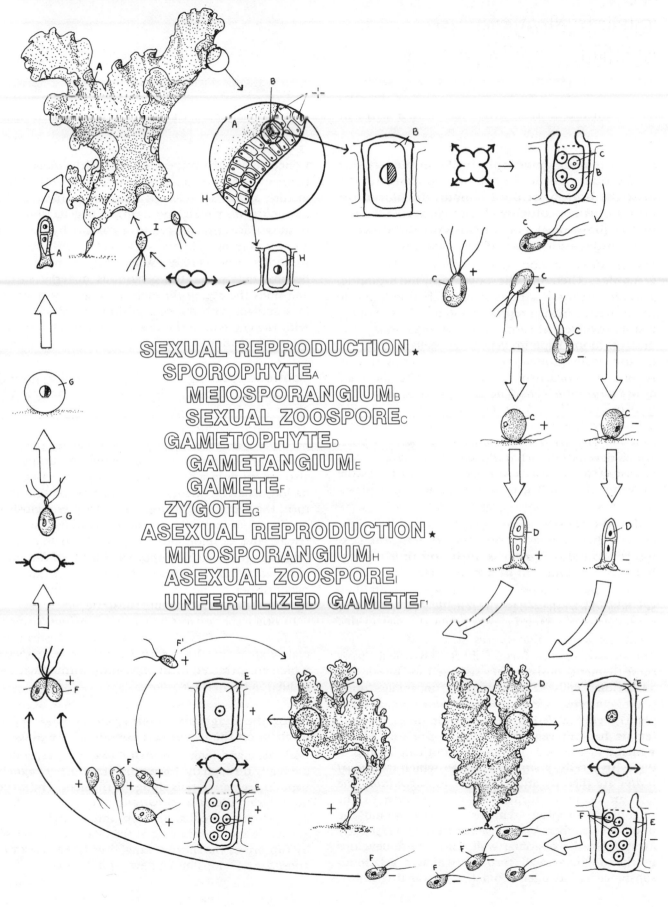

SEXUAL REPRODUCTION ★
SPOROPHYTE A
MEIOSPORANGIUM B
SEXUAL ZOOSPORE C
GAMETOPHYTE D
GAMETANGIUM E
GAMETE F
ZYGOTE G
ASEXUAL REPRODUCTION ★
MITOSPORANGIUM H
ASEXUAL ZOOSPORE I
UNFERTILIZED GAMETE F'

43
GREEN ALGAE (CHLOROPHYTA)

Color the male gametophyte (A) on the left side of the plate, but leave the mitosporangium (L), near the bottom of the filament, uncolored for now. Continue coloring clockwise across the top of the plate until you finish with the diagram illustrating androspore (C) release.

Oedogonium, a common freshwater filamentous green alga, is heterothallic, oogamous, and haplontic. The life history illustrated is one variation found in some species. Sexual reproduction in this type of life history involves the formation of *androsporangia* by one or more haploid vegetative cells within the mature male *gametophyte* filament. A series of mitotic divisions may produce a number of small *androsporangial cells.* The protoplasts of the *androsporangial cells* undergo mitosis to produce special haploid zoospores, called *androspores,* that do not function directly as male gametes, or sperm. A pore, which develops in each *androsporangial cell,* permits release of the enclosed *androspores,* which are motile by a ring of small flagella near their anterior end.

Color the female gametophyte (A) on the right side of the plate, and continue coloring the sexual reproductive process until you finish coloring the first diagram of a zygote (J).

Androspores, attracted to the female *gametophyte* by sex hormones released by the female, attach to one or more haploid vegetative cells along the female filament. Attached *androspores* soon undergo a few mitotic divisions to form a short, few-celled *dwarf male* filament consisting of a stalk cell and one or two antheridial cells. The epiphytic (meaning "upon plant," in reference to its attachment upon another plant) *dwarf male* induces a mitotic division of the female haploid vegetative cell, or *pre–eeg cell,* to which it is attached. The single mitotic division produces two cells, a *support cell,* to which the *dwarf males* are directly attached, and an *oogonium.* The *egg,* or oospore, formed by condensation of the oogonial protoplast, remains in place within the oogonial chamber. As the *egg* matures, it pulls slightly away from the oogonial wall, and a pore develops in the oogonial wall through which a gelatinous substance is extruded from the oogonial chamber.

Concurrently, the antheridial cells of the *dwarf male* filament are producing *sperm* by mitosis. When mature, *sperm* are released from the antheridia. The motile *sperm,* which have a ring of flagella (too small to show), are attracted to the *oogonium* by a sex hormone and trapped by the sticky gelatinous substance the *oogonium* secretes. When one or more *sperm* become entrapped, the gelatinous material shrinks and pulls the *sperm* through the oogonial pore and into contact with the *egg.* Fusion of a single *sperm* with the *egg* within the oogonial chamber results in *zygote* formation.

Continue coloring from the released zygote (J) to the formation of the new gametophyte (A), shown at the bottom right.

The *zygote* produces a thick, resistant cell wall and becomes a dormant zygospore that carries the plant through unfavorable environmental conditions and also functions as a dispersal unit. Just prior to germination, when favorable environmental conditions return, the zygospore protoplast undergoes meiosis to produce four haploid meiospores, two male and two female, that develop into *zoospores* with an anterior ring of flagella. Each *zoospore* is capable of forming a new *gametophyte.* These *zoospores* function as a major dispersal unit and may swim for some distance before settling on a suitable substrate.

In other species of *Oedogonium,* no *androspores* or *dwarf male* filaments are produced. Instead, the male filament produces antheridia that release *sperm,* which are attracted to an *oogonium* formed without induction by an *androspore.*

Color the diagrams illustrating asexual reproduction on the lower left corner of the plate.

Both male and female *gametophytes* may reproduce asexually through the formation of *mitosporangia* by haploid vegetative cells along the filament. The protoplast within the *mitosporangium* condenses to form a single mitospore that develops into a single *asexual zoospore,* motile by a ring of flagella, that is released by rupture of the mitosporangial wall. Each *asexual zoospore* develops into a new *gametophyte.*

OEDOGONIUM LIFE HISTORY.

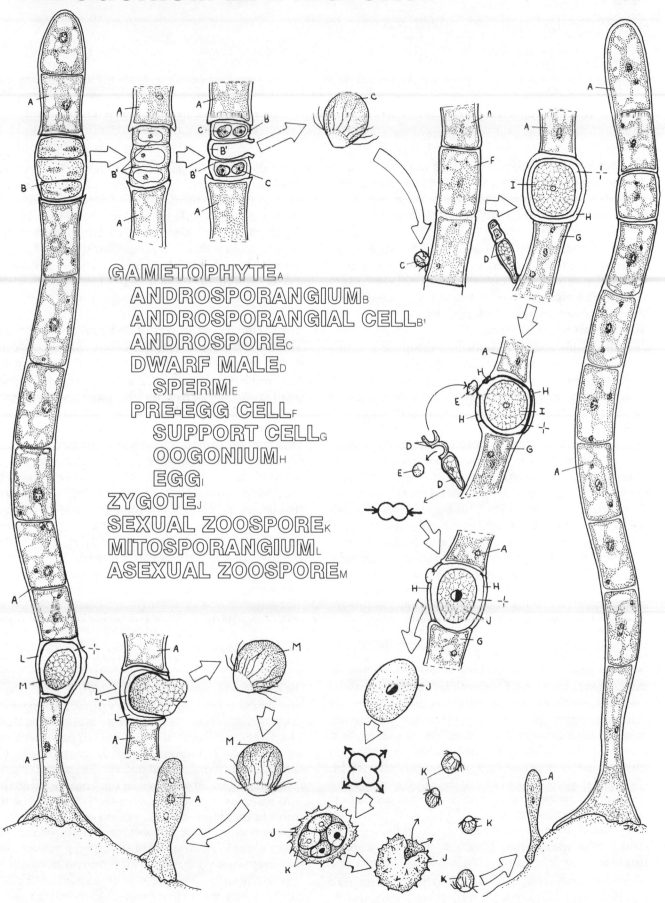

GAMETOPHYTEA
ANDROSPORANGIUMB
ANDROSPORANGIAL CELLB'
ANDROSPOREC
DWARF MALED
SPERME
PRE-EGG CELLF
SUPPORT CELLG
OOGONIUMH
EGGI
ZYGOTEJ
SEXUAL ZOOSPOREK
MITOSPORANGIUML
ASEXUAL ZOOSPOREM

44
GOLDEN-BROWN ALGAE (CHRYSOPHYTA)

The golden-brown algae as treated here include the golden algae, the diatoms, and the yellow-green algae. Both diatoms and the yellow-green algae are distributed in marine and freshwater habitats.

Color the diagram at the top right. Color the diatom labeled "parent cell" at the upper left. Color the sequence depicting asexual reproduction to the right of the "parent cell." Also color the diagrams labeled "large valve sequence" and "small valve sequence".

Diatoms, a unicellular, diploid algal group of over five thousand species, have a modified diplontic life history. Diatoms do not produce a cellulosic cell wall. Instead they produce a two-parted shell consisting of two halves called valves that is composed of up to 95 percent silica or glass. The two halves, one large and one small, overlap like a pillbox. Asexual reproduction is their primary means of reproduction and population increase. During asexual reproduction, each of the two new cells produced receives one old valve from the parent cell. This leaves each new cell with only one valve, or half a shell, and each new *protoplast* must secrete a new valve, within the old valve received from the parent cell, to form a complete two-parted shell. Therefore, each new cell then has one old valve, from the parent cell, and one new valve. Thus, one of the new cells produced following asexual cell division retains the same proportions as the parent cell, while the other, formed within the smaller old valve, is somewhat smaller.

If the fate of the old *larger valve* is followed through successive asexual generations (*generations one, two,* and *three* on the plate), it can be observed that the original parent cell proportions are carried through each generation. If the fate of the smallest valve produced in each generation is followed through successive asexual generations, it can be observed that the overall size of the diatom occupying the *smaller valve* produced each asexual generation continuously diminishes.

Color the diagrams labeled "sexual reproduction."

Sexual reproduction is induced when diatom size reaches a critical minimum. When this occurs, the diploid diatom *protoplast* undergoes meiosis without cell wall formation; so a single cell containing four haploid nuclei is formed. Three of the haploid nuclei abort, leaving a single haploid nucleus within the cell. The haploid cell then leaves its two-parted shell to become a naked haploid *protoplast* that functions as a *gamete.* Fusion of two haploid *gametes* produces a naked diploid *protoplast* that soon secretes a shell having maximum valve dimensions once again.

Color the filaments (I) of the yellow-green alga Vaucheria in the habit sketch. Also color the diagrams of Vaucheria asexual reproduction.

The yellow-green algae, division Xanthophya, is a small group, about four hundred species, of primarily freshwater algae. There are uninucleate, unicellular, or filamentous species and multinucleate species. *Vaucheria,* a common freshwater genus, forms a mass of branched *filaments* that are multinucleate. In asexual reproduction, a cross wall, or *septum,* cuts off the terminal portion of a *filament* to form a *sporangium* that produces a single multinucleate *zoospore* that is motile by numerous flagella. Most other yellow-green algal cells are motile by two unequal (one whiplash and one tinsel) flagella (not shown). The *zoospore* eventually settles on a suitable substrate and germinates to form a new, asexually formed, generation.

Color the diagrams of Vaucheria sexual reproduction.

Sexual reproduction in *Vaucheria* involves the formation of a uninucleate *oogonium* that is separated from the multinucleate *filament* by a *septum.* Each *oogonium* contains one *egg.* The *antheridia,* also separated from the *filament* by a *septum,* are multinucleate. Due to the presence of many nuclei, each *antheridium* produces numerous *sperm* that are motile by two laterally inserted, unequal flagella. *Sperm* are released through one or more pores in the antheridial shell.

At maturity, a pore through which *sperm* may enter forms in the oogonial wall. Fusion of one *sperm* with the *egg* within the *oogonium* forms a *zygote* that produces a thick, resistant wall. Following release from the *oogonium,* the *zygote* may remain dormant for several months. The *zygote* undergoes meiosis to form haploid nuclei just prior to germination.

GOLDEN-BROWN ALGAE INTRODUCTION.

DIATOMS ★
    PARENT VALVES ★
      LARGEA
      NEWA¹
      SMALLB
      NEWB¹
PROTOPLASTC

PROGENY VALVES ★
1st GENERATIOND
2nd GENERATIONE
3rd GENERATIONF
GAMETEG
ZYGOTEH

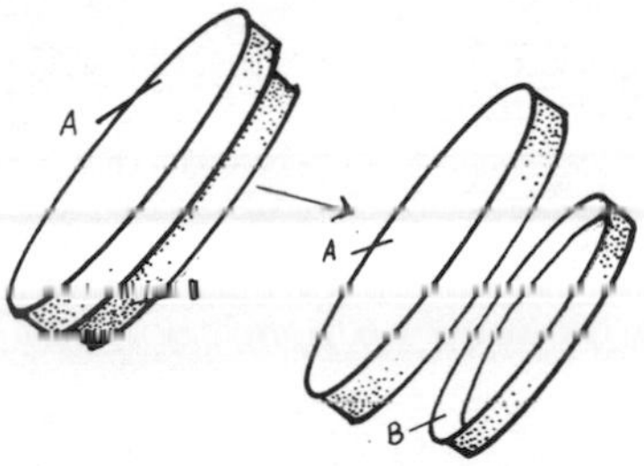

ASEXUAL REPRODUCTION ★

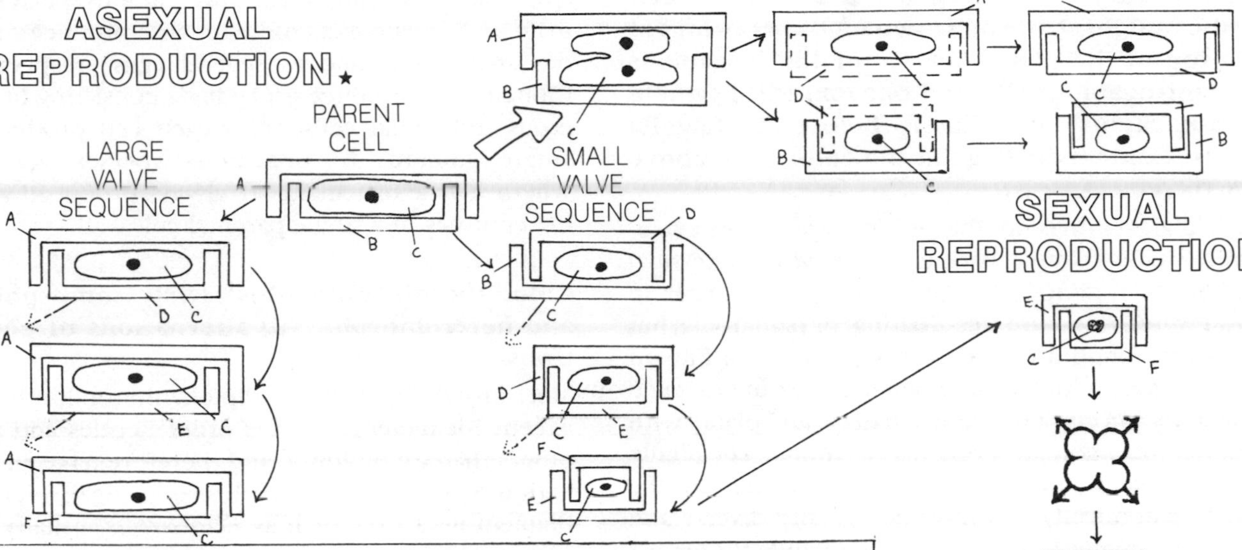

LARGE VALVE SEQUENCE

PARENT CELL

SMALL VALVE SEQUENCE

SEXUAL REPRODUCTION ★

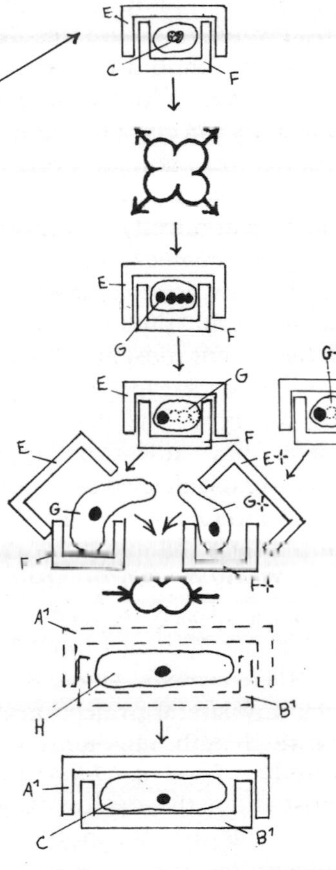

YELLOW-GREEN ALGAE ★
FILAMENTI
SEPTUMJ
SPORANGIUMK
ZOOSPOREL

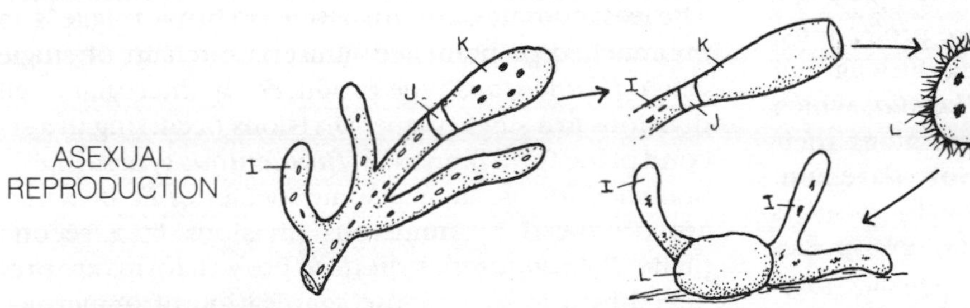

ASEXUAL REPRODUCTION

OOGONIUMM
EGGN
ANTHERIDIUMO
SPERMP
ZYGOTEQ

VAUCHERIA

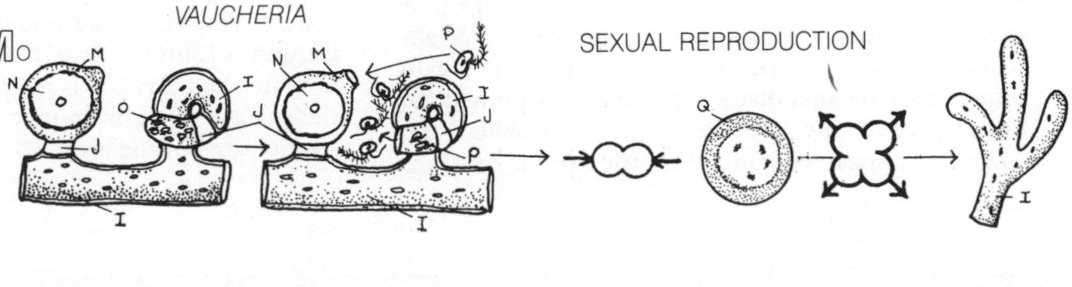

SEXUAL REPRODUCTION

45

BROWN ALGAE (PHAEOPHYTA)

Color over the lines indicating individual tinsel (A) and whiplash (B) flagella, and color the cell bodies a light shade of brown on the drawings at the top of the plate. Also color the isogamous (C), anisogamous (D), and oogamous (E) gamete arrangements immediately below the flagella types. Color their flagella the color used above for flagella (A or B).

Brown algae, division Phaeophyta, are almost exclusively shallow-water marine species of the intertidal and subtidal zones, though a few species occur in fresh water. A few are free-floating, or planktonic, but most are found attached to a substrate. All known species are multicellular. Branched and unbranched filaments, sheets, amorphous masses, and plants with distinct morphological regions are found. The highly complex kelps, or brown seaweeds, which are the most anatomically and morphologically diverse and complex members of the division, include the largest algal plants, approaching 100 meters long. Both diplohaplontic and diplontic life histories occur within the division, but most are diplohaplontic.

Both asexual and sexual zoospores, and at least one gamete type, the male, are motile. Therefore, all species have at least one type of motile reproductive cell. Almost all species have motile cells with two unequal, lateral flagella, one *tinsel* and one *whiplash*, inserted near the middle of the usually kidney-shaped cell. *Tinsel flagella* are distinguished by the presence of rows, usually two, of lateral, hairlike projections, called mastigonemes (not labeled), along the length of the flagellar axis, whereas *whiplash flagella*, which lack any lateral projections, are smooth along their entire length. In most, the anterior flagellum, whether *tinsel* or *whiplash*, though usually *tinsel*, is most often the longest flagellum. Very few species have cells that are motile by a single, apical, *tinsel flagellum*.

Species exhibiting *isogamy, anisogamy,* and *oogamy* are spread throughout the division. Isogamous species produce gametes that are similar in size and shape; anisogamous species produce morphologically similar gametes that distinctly and consistently differ in size, with the male gamete the smaller one. Oogamous species produce motile sperm cells and nonmotile eggs that are conspicuously larger than the sperm. Most isogamous and anisogamous species have male and female gametophytes that produce numerous sperm and one to several eggs, respectively, though some produce gametangia consisting of numerous individual cells, with each cell producing a single gamete. In oogamous species, gametes, whether male or female, are always produced within the confines of a single parental cell wall.

Color the diagrams illustrating isomorphic (F) and heteromorphic (G) alternations of generations.

Most brown algae have a diplohaplontic life history, except for members of the order Fucales and a few others that are diplontic and therefore have an alternation of generations. Simple brown algae, including filamentous forms such as *Ectocarpus*, usually have *isomorphic* generations, whereas more complex forms, including the kelps such as *Laminaria*, have a *heteromorphic* alternation of generations in which the sporophyte, or diploid generation, is a relatively large, conspicuous plant and the gametophyte, or haploid generation, is a minute, microscopic plant.

Color the diagrams illustrating filamentous (H), pseudoparenchymatous (I), and parenchymatous (J) growth forms.

The least complex growth pattern in brown algae is an unbranched or branched, uniseriate (chain of single cells), *filamentous* type produced by intercalary cell divisions in a single plane. Divisions occurring in a second plane form branched, *filamentous* types. Highly branched forms, such as some species of *Ectocarpus*, are produced by numerous divisions in a second plane. *Pseudoparenchymatous* growth forms are produced by a loose or close aggregation of numerous, intertwined, branched filaments that collectively form the plant body. In *Leathesia*, an amorphous brown algal mass is formed. True *parenchymatous* growth is found in brown algae, such as *Nereocystis*, in which cell division is restricted to a layer of cells on the surface that continually adds to the diameter and surface area of the alga.

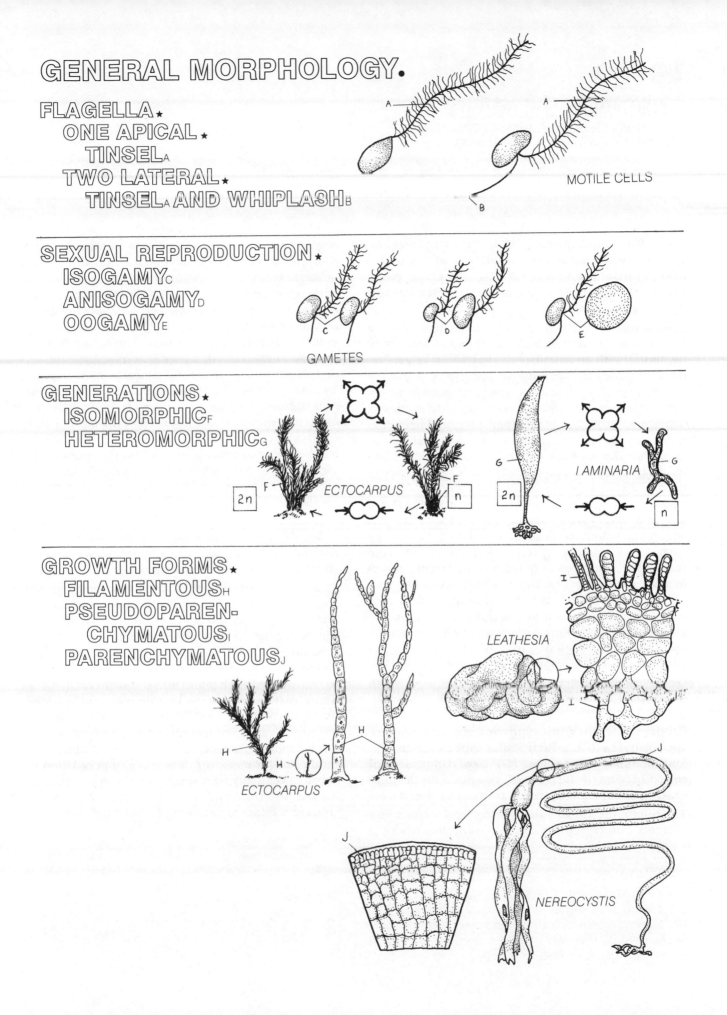

GENERAL MORPHOLOGY.

FLAGELLA ★
   ONE APICAL ★
      TINSELA
   TWO LATERAL ★
      TINSELA **AND WHIPLASH**B

MOTILE CELLS

SEXUAL REPRODUCTION ★
   ISOGAMYC
   ANISOGAMYD
   OOGAMYE

GAMETES

GENERATIONS ★
   ISOMORPHICF
   HETEROMORPHICG

2n   F   *ECTOCARPUS*   F   n

2n   G   *I AMINARIA*   G   n

GROWTH FORMS ★
   FILAMENTOUSH
   PSEUDOPAREN-
      CHYMATOUSI
   PARENCHYMATOUSJ

LEATHESIA

H   H   H

ECTOCARPUS

J

J

NEREOCYSTIS

46

BROWN ALGAE (PHAEOPHYTA)

Color the gametophytes (A), both mating strains, located in the upper left quadrant of the plate. Also, continuing in a clockwise direction, color the plurilocular gametangia (B), gametes (C), and zygote (D).

Ectocarpus, a relatively simple filamentous marine brown alga, is heterothallic, isogamous, and diplo-haplontic with an isomorphic alternation of generations in which the *sporophyte* and *gametophyte* are morphologically similar in size and growth habit. Both generations, diploid and haploid, occupy a prominent position in the life history, and neither is dominant, due to differences in size or duration. The *gametophytes* of both mating strains, plus (+) and minus (−), are morphologically identical. A series of mitotic divisions produces the multicellular *plurilocular gametangia* on short side branches. *Plurilocular gametangia* are so named because they consist of numerous individual cells, each of which produces a single motile *gamete. Gamete* release occurs by rupture of the numerous individual cell walls of a *plurilocular gametangium. Gamete* motility is due to a long, anterior, tinsel flagellum and a shorter, posterior, whiplash flagellum that are laterally inserted on the kidney-shaped cell body. Enormous numbers of motile *gametes* are freely released into the surrounding water where they swim about until they contact a compatible *gamete* of the opposite mating strain. *Gamete* fusion, or syngamy, produces a nonmotile *zygote.*

Color the embryonic sporophyte and mature sporophytes (E), plurilocular sporangium (F), and asexual zoospores (G) depicting asexual reproduction of the diploid, or sporophytic, generation across the extreme bottom of the plate.

The nonmotile *zygote* briefly drifts freely with water currents before settling upon a substrate, where it begins a series of mitotic divisions to form a mature *sporophyte* that has the potential of producing two different types of *sporangia.* In asexual reproduction of the *sporophyte,* multicellular *plurilocular sporangia,* which closely resemble the *plurilocular gametangia* produced by *gametophytes* though the individual cells formed are diploid rather than haploid, provide a means of rapid population increase and further dispersal for the diploid generation. As in *plurilocular gametangium* formation, a *plurilocular sporangium* consists of numerous individual cells formed by mitotic divisions. Each of the diploid mitosporangial cells of the *plurilocular sporangium* releases a single *asexual zoospore* that may disperse for some distance before settling on a suitable substrate, where it undergoes mitotic divisions to produce another *sporophyte.*

Color the unilocular sporangium (H) and sexual zoospores (I), as well as the previously listed structures on the remaining diagrams.

A second *sporangium* type, the unicellular *unilocular sporangium,* is a component of the sexual cycle. *Unilocular sporangia* develop as enlarged cells, containing a single protoplast, that form on the tips of specialized side branches. The single diploid nucleus within the *unilocular sporangium* undergoes meiosis to produce four haploid nuclei, two of each mating strain, that then undergo one or more mitotic divisions to produce several haploid meiospores that are released as *sexual zoospores* upon rupture of the unilocular sporangial wall. *Sexual zoospores* function as another major means of dispersal and population increase. Eventually, the *sexual zoospores* settle on a suitable substrate and undergo mitotic divisions to produce another generation of *gametophytes.*

ECTOCARPUS LIFE HISTORY.

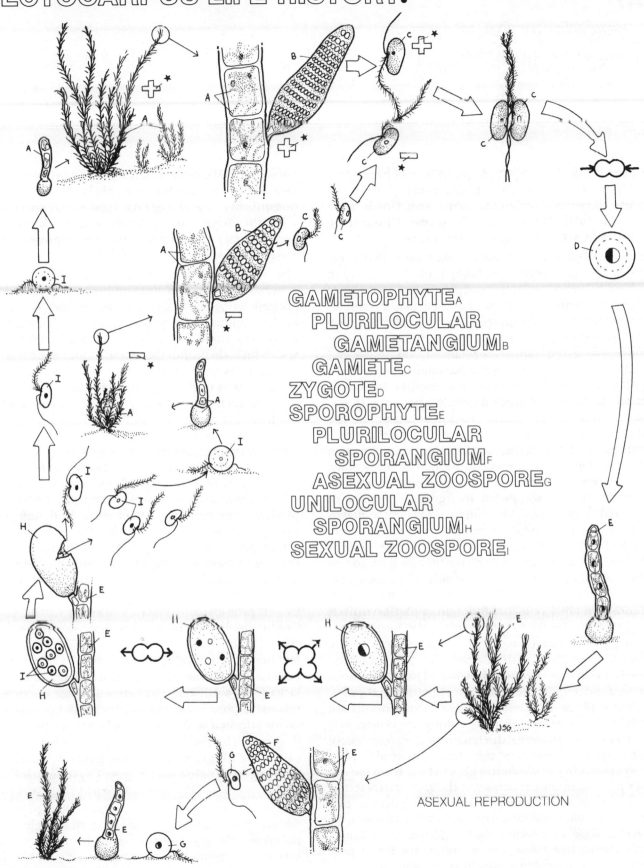

GAMETOPHYTE A
PLURILOCULAR GAMETANGIUM B
GAMETE C
ZYGOTE D
SPOROPHYTE E
PLURILOCULAR SPORANGIUM F
ASEXUAL ZOOSPORE G
UNILOCULAR SPORANGIUM H
SEXUAL ZOOSPORE I

ASEXUAL REPRODUCTION

47

BROWN ALGAE (PHAEOPHYTA)

Color the holdfast (A¹), stipe (A²), and blade (A³) of the sporophyte (A) on the right side of the plate. Continue coloring until you finish with the medulla (F) on the diagrams illustrating details of stipe anatomical structure.

Laminaria is a heterothallic, oogamous kelp that exhibits a heteromorphic diplohaplontic life history in which a dominant, macroscopic (large) *sporophyte* alternates with a cryptic, microscopic (small) gametophyte that shares no morphological similarities with the *sporophyte*. The *sporophyte* consists of one or more elongated *blades*, or lamina, up to several meters in length, with narrowed stalklike *stipes* and a *holdfast* composed of numerous rootlike structures called haptera (not labeled), which hold the *sporophyte* to a solid, usually rocky, substrate in the shallow waters of the intertidal and subtidal zones.

Typical of the kelps, *Laminaria* exhibits a high degree of anatomical specialization. The plant body, completely covered by a *mucilaginous layer*, has four distinct, though somewhat intergrading, regions of internal stem tissues. The outermost layer, the *meristoderm,* is composed of numerous, small, closely packed cells that form an epidermislike layer. Cells of the *meristoderm* are continually dividing to add to the girth of the *sporophyte*. The bulk of the *stipe* consists of the *cortex*, which is made of thin-walled, parenchymalike cells that function in photosynthesis and storage. Internal to the *cortex*, the isodiametric (having the same diameter throughout) cells of the *cortex* intergrade with filamentous cells from the *medulla* in the *transition zone*. The filamentous *medulla* occupies the central portion of the *stipe*.

Many of the filamentous cells of the *medulla* are oriented in a vertical direction along the central axis to form long filaments the length of the *stipe*. These are adapted to carry food materials, produced by photosynthesis in the *blade* and upper *stipe*, from the upper photosynthetic portions of the *sporophyte* to the lower portions. The lower parts are often deeper than effective light penetration of the wavelengths used by brown algae in photosynthesis. Therefore, without this conductive tissue, the *medulla*, the lower portions of the *sporophyte* would starve and die.

Starting at the top left, color the blade (A³) enlargement, the meiosporangium (G), sexual

zoospores (H), male and female gametophytes (I) and (I¹) respectively, antheridia (J), sperm (K), oogonium (L), and egg (M). Stop when you finish with the diagrams depicting sperm (K) release and egg (M) extrusion. Note the counterclockwise progression of the life history.

The sexual reproductive structures of the diploid generation, the *meiosporangia*, develop within pockets along the margins of the *blade*. The single nucleus within each meiosporangial cell undergoes meiosis to produce four haploid nuclei, two of each sex, which then undergo one or more synchronous mitotic divisions to form several meiospores that are released as *sexual zoospores* through a pore in the apex of each *meiosporangium*. *Sexual zoospores*, motile by a long anterior tinsel flagellum and a shorter posterior whiplash flagellum inserted laterally in the kidney-shaped cell body, are a major means of dispersal and population increase for *Laminaria*.

After some time, the *sexual zoospores* settle upon a substrate and undergo a series of mitotic divisions to produce gametophytes, either *male* or *female*, consisting of few to several cells. The two types of gametophytes differ somewhat from one another. The *male gametophyte*, a microscopic branched filament, forms *antheridia* from the terminal cells of the branches. A series of mitotic divisions of the antheridial cell protoplast produces numerous motile *sperm* that resemble, but are smaller than, the *sexual zoospores*. The *female gametophyte* produces one to a few *oogonia* through transformation of one or more cells, and each *oogonium* produces a single *egg* that lacks a cell wall. Upon maturity, the *egg* is extruded through a pore in the oogonial wall, but it usualy remains attached to the outside of the *oogonium*.

Color the diagrams depicting gamete, sperm (K) and egg (M), fusion and zygote (N) formation and development of the new sporophyte generation.

Sperm, attracted by a sex hormone secreted by the *female,* gather about the *egg* cell until a single *sperm* penetrates the *egg* and fuses with it to effect fertilization, or syngamy, which produces a *zygote* to start a new diploid generation. The *zygote* floats briefly at the mercy of water currents until it attaches to a suitable substrate, where it germinates by mitosis to begin the start of a new *sporophyte*.

LAMINARIA LIFE HISTORY.

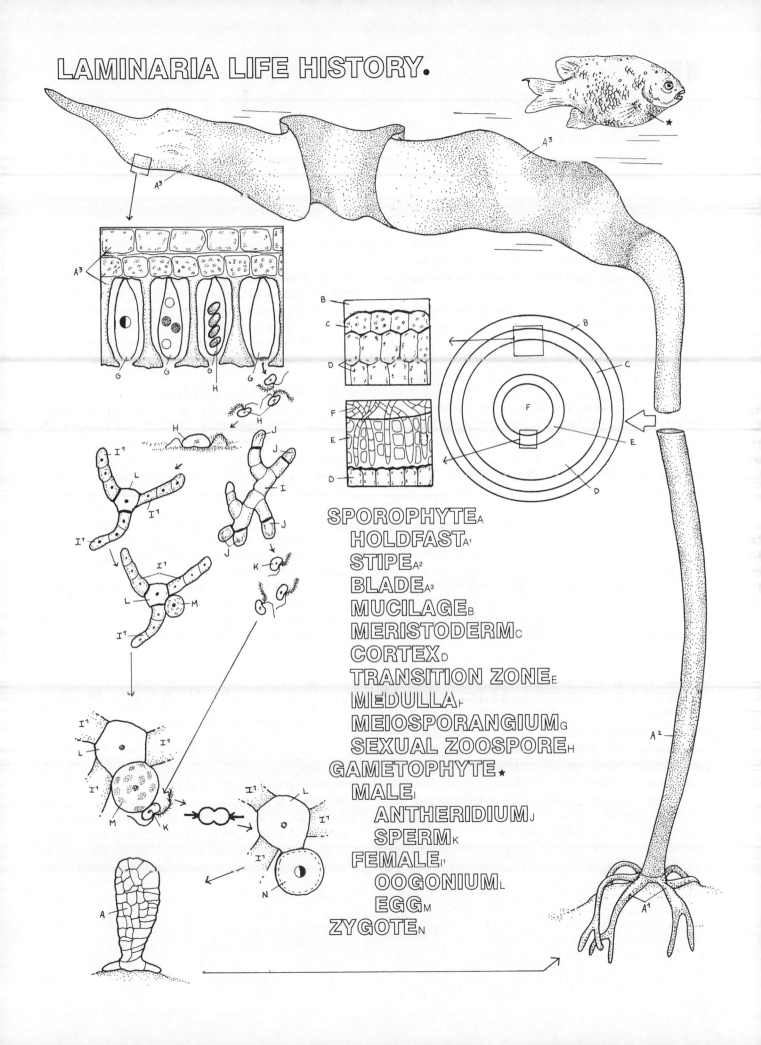

SPOROPHYTE A
HOLDFAST A1
STIPE A2
BLADE A3
MUCILAGE B
MERISTODERM C
CORTEX D
TRANSITION ZONE E
MEDULLA F
MEIOSPORANGIUM G
SEXUAL ZOOSPORE H
GAMETOPHYTE ★
MALE I
ANTHERIDIUM J
SPERM K
FEMALE I1
OOGONIUM L
EGG M
ZYGOTE N

48
BROWN ALGAE (PHAEOPHYCOPHYTA)

Color the sporophyte plant (A) in the middle of the plate and the conceptacles (B), shown enlarged out of proportion, located on the tips of the branches.

Fucus, or rockweed, is a common temperate region marine brown alga of the kelp group that is homothallic, oogamous, and diplontic. The dominant diploid phase, or *sporophyte,* is found growing attached to rocks in the intertidal zone. The *sporophyte* is dichotomously branched, having branches that form a Y shape. The plant has laminar, or flattened, thick, leathery branches that become broadened toward their tips. Under favorable conditions of light and temperature, mature *sporophytes* develop swollen branch tips that will produce sexual reproductive structures known as *conceptacles.* Other swollen areas, gas bladders, which may be located on the branches, contain gases that function to keep the plant erect in the water.

During sexual reproduction, numerous roughly spherical, hollow *conceptacles,* appearing as raised tubercles, form embedded within the enlarged branch tips. Gametangia are produced within the *conceptacle* chambers, which contain either male or female gametangia or both, depending on species.

Color the male conceptacle (B) and its associated structures at top left and continue coloring down the left margin until you finish with the diagram illustrating sperm (F) release. Note that a species with unisexual (only one sex present) conceptacles is illustrated and that the conceptacle wall consists of numerous compactly aggregated cells.

Male *conceptacles* contain unbranched, sterile, diploid filaments, called *paraphyses,* that extend into the central portion of the *conceptacle* chamber and branched, diploid, *fertile filaments* that produce swollen *antheridia* at their tips. Meiosis within each single-celled *antheridium* produces four haploid male nuclei that then undergo a series of synchronous mitotic divisions to produce numerous *sperm.* When mature, the biflagellated *sperm,* with an anterior tinsel flagellum and a trailing whiplash flagellum, are released into the male *conceptacle* chamber, and an enormous number of *sperm* are freely dispersed into the surrounding water from each *conceptacle.*

Color the female conceptacle (B) and its associated structures at top right and continue coloring down the right margin until you finish with the diagram illustrating egg (H) release.

Female *conceptacles* also contain unbranched *paraphyses,* but the diploid *fertile filaments* consist of a single stalk cell and a single, terminal *oogonium.* Each single-celled *oogonium* undergoes meiosis to produce four haploid female nuclei followed by a single mitotic division to produce a total of eight *eggs* that are nonmotile and considerably larger than the *sperm.* At maturity and under environmental conditions compatible with those required for *sperm* release, the eight *eggs,* which lack a cell wall, are released into the female *conceptacle* chamber and, through its pore, into the surrounding water, where dispersal is at the mercy of water currents.

Color the diagram illustrating union of sperm (F) and egg (H) at the bottom and continue coloring until you finish with the plate.

The nonmotile *eggs* secrete a sex hormone that attracts the motile *sperm,* which cluster around the *egg.* Eventually, a single *sperm* fuses with the *egg* to form a *zygote,* which soon forms a cell wall and secretes a sticky substance on its exterior surface that will enable it to adhere to a substrate. The *zygote* floats freely for a time and is a major unit of dispersal, but it soon settles and begins growth to form a new *sporophyte* that is securely anchored to the substrate by a holdfast.

FUCUS LIFE HISTORY.

SPOROPHYTE ᴀ
CONCEPTACLE ʙ
PARAPHYSES ᴄ
FERTILE FILAMENT ᴅ
GAMETANGIUM ★
ANTHERIDIUM ᴇ
SPERM ꜰ

OOGONIUM ɢ
EGG ʜ
ZYGOTE ɪ

49
RED ALGAE (RHODOPHYTA)

On the diagrams labeled "Bangiophycidae," color the gametophyte (A), vegetative cell (A¹), spermatangium (B), spermatia (C), carpogonium (D), and egg (E).

The red algae, Rhodophyta, lack motile cells. Their structure ranges from uninucleate, unicellular types to large, complex, multicellular forms with multinucleate cells. All are oogamous, and most have a rather complex life history. Most are diplohaplontic, but some have a second diploid generation that is attached to the female *gametophyte.* The class Rhodophyceae is divided into two subclasses, the Bangiophyciade and Florideophyciade, primarily based upon reproductive differences.

The Bangiophycidae is a small group consisting of fewer than one hundred species. Unicellular, filamentous, and sheetlike species occur. The gametangia are derived from *vegetative cells* and are relatively simple. The male gametangium, called a *spermatangium,* consists of a single *vegetative cell* that undergoes repeated mitotic divisions to produce numerous male gametes, called *spermatia.* The nonmotile *spermatia* are released into the open water by rupture of the spermatangial wall. Dispersal is by water currents. Like the male gametangium, the female gametangium, called a *carpogonium,* is formed by a *vegetative cell,* but no mitotic divisions occur. Instead, the protoplast of the *vegetative cell* differentiates to become an *egg.* A small portion of the *egg,* the *trichogyne* (not separately colored), protrudes through the carpogonium wall (formerly the *vegetative cell* wall) as a sticky structure receptive to *spermatia.*

Color the previously listed features and the trichogyne (F) on the diagrams labeled "Florideophycidae."

The reproductive structures and life histories of members of subclass Florideophycidae are more complex than those of the Bangiophycidae. *Spermatangia* are usually formed in clusters on a specialized branch, the *spermatangial branch.* Each *spermatangium* produces only one *spermatium. Carpogonia* are also usually formed on a specialized branch as specialized cells (unlike the Bangiophycidae). The *carpogonium* usually has a long, fingerlike *trichogyne.*

Starting with the gametophyte below "generations," color the previously listed features and the pericarp (G), zygote (H), and carposporophyte (I). Stop when you finish with the first diagram depicting a carposporophyte. Choose a light color for the pericarp.

The diagrams labeled "generations" introduce the features and complexity of Florideophycidae life histories. Species are typically heterothallic, having separate male and female *gametophytes,* but both *gametophyte* types are similar in form. The *carpogonia* produced by the female *gametophytes* are fertilized by free-floating *spermatia* released by male *gametophytes.* This union results in the formation of a *zygote.* In many red algae, fertilization apparently stimulates the female *gametophyte* to produce a protective, flask-shaped structure, called a *pericarp,* that surrounds the *zygote.* Mitotic division of the *zygote* produces a few to many-celled sporophyte called a *carposporophyte* within the *pericarp.*

Now color the meiosis symbol, previously listed features, and haploid carpospores (J) on the path leading to the left and labeled "two generations." Do not color the components of the pathway labeled "three generations" yet.

Depending on species, a complete life history involves two generations, one gametophytic and one sporophytic, or three generations, one gametophytic and two sporophytic. In the two-generation life history, some or all cells of the diploid *carposporophyte* undergo meiosis to produce meiospores called *carpospores.* Characteristic of the two-generation life history, *haploid carpospores* are produced, as they are the direct product of meiosis. These *haploid carpospores* are dispersed by water currents. If they land on a suitable substrate, *haploid carpospores* germinate to produce a new *gametophyte.*

Now color the components of the pathway labeled "three generations," including the diploid carpospores (K), tetrasporophyte (L), and tetraspores (M).

In the three-generation life history, cells of the *carposporophyte,* the first sporophytic generation, do not undergo meiosis to produce *haploid carpospores.* Instead, the diploid *carposporophyte* produces *diploid carpospores* by mitosis. Like *haploid carpospores,* the *diploid carpospores* are released to function as major dispersal units. When a *diploid carpospore* germinates, it produces the second sporophytic generation, the *tetrasporophyte,* which in most red algae is similar to the *gametophyte* in form. Diploid cells within the *tetrasporophyte* undergo meiosis to produce meiospores called *tetraspores.* Release of *tetraspores* is through rupture of the *tetrasporophyte.* Like the *diploid carpospores,* the *tetraspores* are freely dispersed by water currents. If they land on a suitable substrate, *tetraspores* germinate to produce a new *gametophyte,* thus completing the life history.

RED ALGAE INTRODUCTION.

GAMETOPHYTE_A
VEGETATIVE CELL_{A¹}
GAMETANGIA ★
SPERMATANGIUM_B
SPERMATANGIAL
BRANCH_{B¹}
SPERMATIUM_C
CARPOGONIUM_D
EGG_E
TRICHOGYNE_F

BANGIOPHYCIDAE

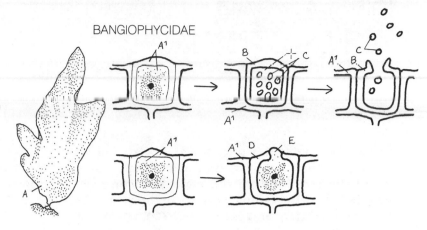

FLORIDEOPHYCIDAE

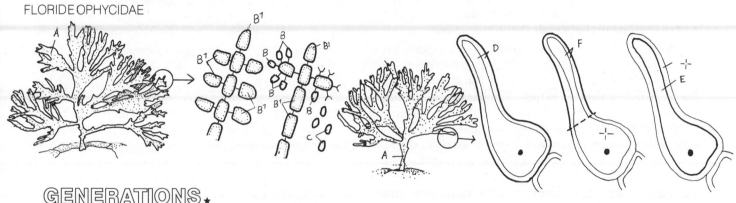

GENERATIONS ★

PERICARP_G
ZYGOTE_H
SPOROPHYTE ★
CARPOSPOROPHYTE_I
HAPLOID CARPOSPORE_J
DIPLOID CARPOSPORE_K
TETRASPOROPHYTE_L
TETRASPORE_M

TWO GENERATIONS

OR

THREE GENERATIONS

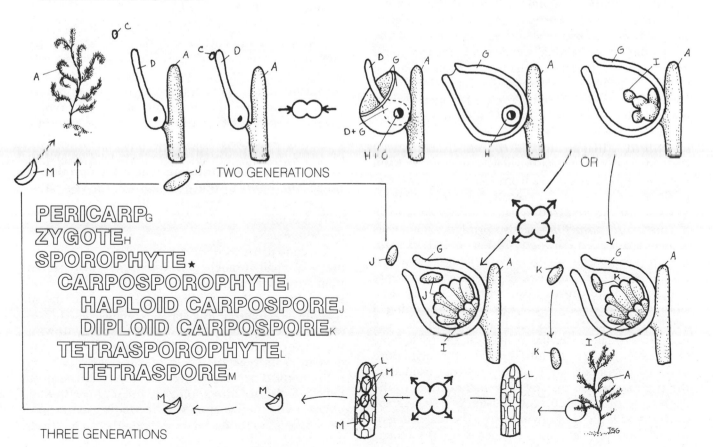

50
RED ALGAE (RHODOPHYTA)

Color the male (A¹) and female (A²) gameto-phytes on the upper left half of the plate. Also, color the close-ups to the right depicting the vegetative cells from each gametophyte, the spermatangia (B), and carpogonia (D), which are male and female gametangia, the spermatia (C), the egg (E), and zygote (F). Stop coloring when you encounter the symbol for meiosis.

Porphyra, subclass Bangiophycidae, is a heterothallic, oogamous, marine red alga having a haplontic life history that may have three separate and morphologically different haploid stages. The gamete-producing stage, or *gametophyte,* consists of sheetlike foliose *male* and *female gametophytes* that closely resemble one another vegetatively but differ in kinds of sexual structures formed. Areas of gametangia formation in both are distinguished by their pale, almost colorless, coloration in contrast with the darker red areas that remain vegetative. Though all cells of the *gametophyte* may become reproductive, the usual progression is from the margins inward, with most of the *gametophyte* remaining vegetative. In the *male gametophyte,* vegetative cells near the margin of the foliose blade first become reproductive as they differentiate into *male* gametangia, called *spermatangia,* that, through a series of mitotic divisions, produce numerous, pale-colored, nonmotile sperm called *spermatia.* When mature, the *spermatia* are released through a pore in the outer spermatangial wall. Each *spermatangium* produces from 16 to 128 *spermatia,* depending on species. As a result, enormous numbers of *spermatia* are released by each *male gametophyte.*

The *female gametophyte* also produces numerous gametangia, called *carpogonia,* but, unlike *spermatangium* formation, a vegetative cell entering sexual reproduction on the *female gametophyte* simply undergoes a transformation, without mitotic division, to form a single *carpogonium* containing a single *egg* that remains in place. When the *egg* matures, a small fingerlike projection, the trichogyne (not labeled), protrudes into the water through a pore in the carpogonial wall. A sticky substance exuded by the trichogyne entraps any *spermatia* carried to the trichogyne by water currents. Fusion of a single *spermatium* with the trichogyne allows the male nucleus to pass into the *egg* and to fuse with the female nucleus to form a *zygote* that remains in place within the *carpogonium.*

Color the diagrams showing carposporangium (G) formation and carpospore (H) release.

The *carpogonium* is transformed into a *carposporangium* when meiosis, which apparently soon follows *zygote* formation, produces four haploid nuclei, two of each sex, that may or may not undergo a series of mitotic divisions, depending on species. If no mitotic division occurs, the *carposporangium* produces four *carpospores;* but most species undergo mitosis to produce from eight to sixty-four *carpospores,* which, when mature, are released through the pore in the carposporangial wall through which the trichogyne protruded. Thus, the sequence of metamorphic changes is from vegetative cell to *carpogonium* to *carposporangium.*

Color the diagrams along the bottom of the plate showing the various potential pathways for carpospore (H) development.

When the *carpospores,* which are released and freely dispersed by water currents, land on a substrate, they may follow any one of three developmental pathways. Under certain conditions, a *carpospore* may germinate and begin development into a foliose *gametophyte,* but development of the embryonic *gametophyte* may become arrested after a few mitotic divisons when it consists of few to several cells. One or more vegetative cells may then become transformed into *monosporangia,* which each produce a single *monospore* by transformation of the vegetative protoplast. Under favorable conditions, a *monospore* will germinate and develop into a foliose *gametophyte.*

A second pathway of *carpospore* development produces a small, branched *filamentous plant,* often found growing on the calcareous shells of dead mollusks, that is called the Conchocelis stage because it was once recognized as a separate genus called *Conchocelis.* As in the previously described pathway, *monosporangia* that produce *monospores* may develop from vegetative cells of this stage.

In the third developmental pathway, a *carpospore* develops directly into a foliose *gametophyte.* In *Porphyra,* two types of spores, *carpospores* and *monospores,* account for dispersal and population increase.

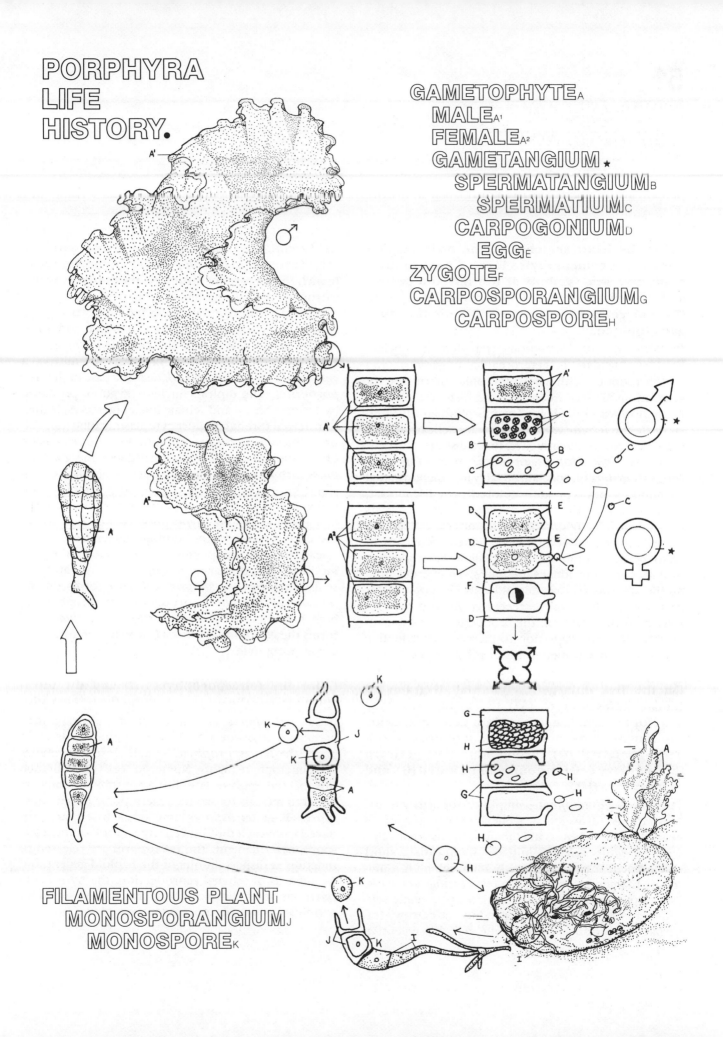

PORPHYRA LIFE HISTORY.

GAMETOPHYTEA
MALEA1
FEMALEA2
GAMETANGIUM★
SPERMATANGIUMB
SPERMATIUMC
CARPOGONIUMD
EGGE
ZYGOTEF
CARPOSPORANGIUMG
CARPOSPOREH

FILAMENTOUS PLANT
MONOSPORANGIUMJ
MONOSPOREK

51
RED ALGAE (RHODOPHYTA)

Color the habit sketches of the male (A) and female (A¹) gametophytes located next to their respective sex symbols at top left and center, and the enlargements of spermatangial clusters (B) and released spermatia (B¹) on the male gametophyte.

Polysiphonia, a filamentous red alga in subclass Florideophycidae, is heterothallic, oogamous, and diplohaplontic, with an isomorphic alternation of generations. It has a complex life history typical of many floridean red algae. The morphologically similar *male* and *female gametophytes* are delicate, highly branched, and grow attached to a substrate.

The *male gametophyte* produces compact *spermatangial clusters* on specialized lateral branches, called trichoblasts (not separately colored). The trichoblast, the main axis of the *spermatangial cluster,* consists of a chain of single cells, or a uniseriate filament, unlike the vegetative branches, which are chains of many cells, or a multiseriate filament. Each *spermatangial cluster* produces (by mitosis) numerous nonmotile *spermatia* that depend entirely upon water currents for dispersal to receptive *female gametophytes.* As in *Porphyra,* the production of numerous *spermatia* is an adaptation to increase reproductive efficiency, since gamete transfer is by chance.

On the five enlarged sequential diagrams of female gametophyte (A¹) to carposporophyte development in the lower right quadrant, color the female gametophyte (A¹), support cell (C), carpogonial cell (D), carpogonium (D¹), pericarp (E), auxillary cell (F), gonimoblast initial (G), and carpospores (H).

The *female gametophyte,* simplified for clarity in the illustrations, forms female gametangia near the tips of vegetative branches. Gametangium formation begins with a *support cell* that initiates (by mitosis) a short side branch, the carpogonial branch, which is a uniseriate filament consisting of a few *carpogonial cells.* The carpogonial branch is terminated by a single cell, called a *carpogonium,* that produces a narrow, fingerlike projection, the trichogyne, which is receptive

to *spermatia.* As the carpogonial branch develops, cells surrounding the *support cell* in the vegetative branch form a vaselike structure, the *pericarp,* which surrounds the carpogonial branch.

When a *spermatium* contacts the sticky trichogyne surface of a *carpogonium,* it adheres to it and transfers its male nucleus into the *carpogonium.* At the same time, the *support cell* produces (by mitosis) a cell, called the *auxillary cell,* near the base of the *carpogonium.* The diploid nucleus (zygote), produced by fusion of male and female nuclei (gametes) is then transferred through a short cytoplasmic tube (not colored) into the *auxillary cell.* A single mitotic division of the diploid *auxillary cell* produces the *gonimoblast initial,* which, through a series of mitotic divisions, produces the diploid carposporophyte (not shown). Almost all the cells of the carposporophyte may become diploid *carpospores,* which function as dispersal units for the diploid generation. Upon release through an opening in the *pericarp,* the nonmotile *carpospores* are freely dispersed by water currents. In review, a *support cell* that forms on the *female gametophyte* produces the carpogonial branch and an *auxillary cell,* and the *auxillary cell* forms the *gonimoblast initial,* which then forms the carposporophyte.

Color the tetrasporophytes (I) and the tetraspores (I¹) on the diagrams along the left margin.

If the *carpospores* land on a suitable substrate, they germinate to form a free-living diploid generation plant, the *tetrasporophyte,* which closely resembles the gametophyte plants. Numerous cells in the branch tips of the multiseriate tetrasporophyte filaments undergo meiosis to produce meiospores called *tetraspores.* Four *tetraspores,* two of each sex, are produced from each diploid cell. By rupture of the tetrasporophyte filament, the *tetraspores* are released to function as dispersal units of the haploid generation. As a means of asexual reproduction, the *tetrasporophyte* can form diploid mitospores (not shown) that germinate and form more *tetrasporophytes.*

POLYSIPHONIA LIFE HISTORY.

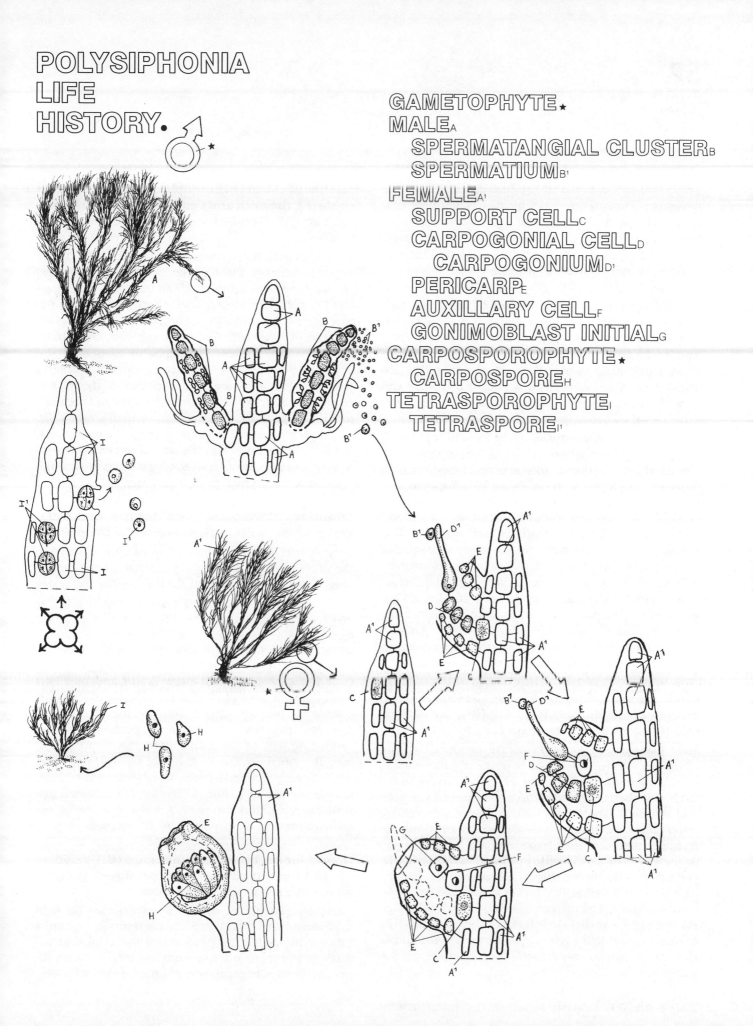

GAMETOPHYTE ★
MALE A
   SPERMATANGIAL CLUSTER B
   SPERMATIUM B¹
FEMALE A¹
   SUPPORT CELL C
   CARPOGONIAL CELL D
    CARPOGONIUM D¹
   PERICARP E
   AUXILLARY CELL F
   GONIMOBLAST INITIAL G
CARPOSPOROPHYTE ★
CARPOSPORE H
TETRASPOROPHYTE I
   TETRASPORE I¹

The bryophytes, division Bryophyta, consist of the liverworts, hornworts, and mosses. They lack true roots, stems, and leaves, though analogous, less-specialized structures are found in many mosses. Most are small plants growing appressed to a continually or frequently moist surface (substrate).

Beginning in the top left corner of the plate, color the meiospores (A), male and female gametophytes (vegetative) (B), antheridiophore (B¹), and archegoniophore (B²), but do not color the boxed gemmae cups (B³) and gemmae (J) illustrated on the vegetative gametophytes or the boxed asexually produced gametophytes (B⁴). Stop before coloring the mitosis symbols.

The liverwort *Marchantia,* class Hepaticae, like all bryophytes is homosporous and diplohaplontic, with a heteromorphic alternation of generations in which there is an independent, prevalent gametophyte generation and a dependent, transient sporophyte generation. The germination of a *meiospore* in a favorable, moist environment initiates formation of a flattened, prostrate *gametophyte* that grows appressed to a moist substrate. Under suitable conditions, mature *gametophytes* produce sexual reproductive structures. *Marchantia* has separate male and female *gametophytes* that produce stalked, gametangia-bearing structures called *antheridiophores* (male) and *archegoniophores* (female) that elevate the mature gametangia well above the surface of the *gametophyte.*

Color the right mitosis symbol, the longitudinal section of the antheridiophore (B¹), the antheridia (C), and the released sperm (D).

Male gametangia, *antheridia,* are produced within chambers on the upper surface of the disclike portion of an *antheridiophore*, which resembles an umbrella. Each chamber is open to the atmosphere by a pore (not colored). The *antheridia* consist of a protective outer jacket, made up of a single layer of cells, and the fertile, gamete-forming tissue within. Mitotic division of the enclosed fertile cells produces numerous biflagellated motile *sperm. Sperm* release occurs through rupture of the antheridial jacket when external moisture is present. The mature *sperm* then swim through the pore in the antheridial chamber and through a film of water covering the *gametophytes* toward the female gametangia in the *archegoniophores* on the female *gametophyte.*

Color the left mitosis symbol, the longitudinal

section of the archegoniophore (B²), and archegonia (E), the neck and ventral canal cells (F), and the egg (G). Stop before coloring the syngamy symbol.

Female gametangia, *archegonia,* are produced on the lower surface of the disclike portion of an *archegoniophore,* which resembles an umbrella lacking its cloth cover. Many *archegonia* may be found near the base of each spokelike projection close to the supporting stalk. The venter, or swollen egg-containing base, of each *archegonium* is deeply imbedded in the *archegoniophore;* the long neck portion projects downward. While immature, the apex of the neck is covered by four cover cells (not shown), and the neck canal is plugged by the *neck* and *ventral canal cells.* As the single *egg* within each *archegonium* matures, the cover cells open and the *neck* and *ventral canal cells* degenerate, leaving an open passageway through which the *sperm* may pass to the *egg.*

Continue coloring until you finish with the diagrams at the bottom right corner of the plate.

Syngamy results in the formation of a *zygote* within the *archegonium.* Through the *gametophyte,* the *archegonium* provides for all the nutritional and water requirements for the developing *sporophyte,* which is totally dependent, or parasitic, upon the *gametophyte.* The enlarged base of the developing *sporophyte,* the *foot,* remains imbedded in the *gametophyte* tissue of the *archegonium* and functions to anchor the *sporophyte* and draw water and nutrients from the *gametophyte.* The *seta,* or stalk, of the *sporophyte* is initially short, but as the *sporophyte* matures, the *seta* elongates to push the spore-producing part of the *sporophyte,* the *sporangium,* into the open to allow dispersal of the numerous *meiospores* produced by meiosis of sporogenous cells within the *sporangium.* At maturity, the sporangial wall ruptures, and the *meiospores* drop free of the *sporangium,* to be dispersed by air currents.

Now color the boxed gemmae cups (B³), gemmae (J), and the boxed asexually produced gametophyte (B⁴) at the top of the plate.

Asexual reproduction in *Marchantia* occurs through the formation of specialized asexual reproductive units called *gemmae* that are produced within cuplike structures, *gemmae cups.* These multicellular asexual reproductive units may produce additional *gametophytes.*

MARCHANTIA LIFE HISTORY.

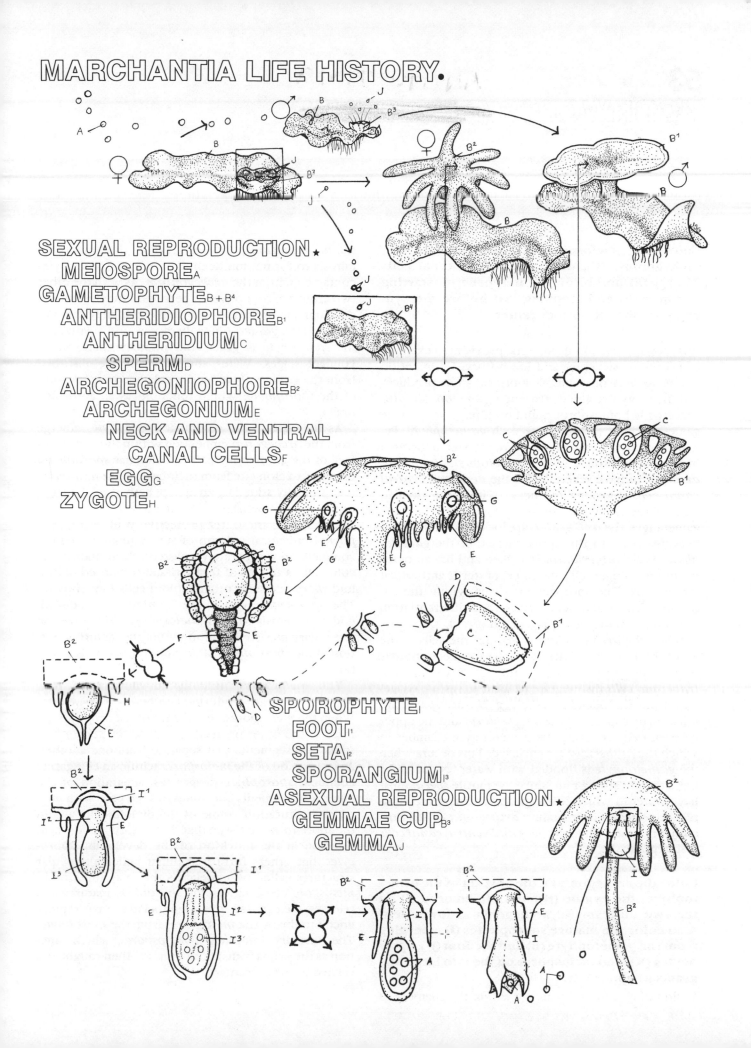

SEXUAL REPRODUCTION ★
- MEIOSPORE A
- GAMETOPHYTE B + B⁴
- ANTHERIDIOPHORE B¹
- ANTHERIDIUM C
- SPERM D
- ARCHEGONIOPHORE B²
- ARCHEGONIUM E
- NECK AND VENTRAL CANAL CELLS F
- EGG G
- ZYGOTE H

SPOROPHYTE I
- FOOT I¹
- SETA I²
- SPORANGIUM I³

ASEXUAL REPRODUCTION ★
- GEMMAE CUP B³
- GEMMA J

Color the meiospore (A), gametophyte (B), archegonium (C), neck and ventral canal cells (D), egg (E), antheridia (F), and sperm (G) starting at top right and stopping just before the syngamy symbol at bottom center.

Anthoceros, division Bryophyta, provides an example of hornwort structure and life history. Germination of a *meiospore* on a suitable moist substrate produces a thin, wavy-margined, flattened *gametophyte. Anthoceros* is homosporous, and both male and female gametangia are produced by each *gametophyte.* Female gametangia, the *archegonia,* are well imbedded in the upper surface of the *gametophyte.* Immature *archegonia* have several cells, the *neck and ventral canal cells,* blocking the neck. Within the swollen base of the *archegonium* is a single *egg.* The male gametangia, the *antheridia,* are located within chambers just beneath the upper surface of the *gametophyte.* Each *antheridium* is stalked and has an outer covering, a single cell layer thick, of sterile antheridial jacket cells. The sporogenous tissue within the antheridial jacket undergoes mitosis and differentiation to produce numerous biflagellated *sperm.*

When the *archegonium* is mature, the cells at the tip of the archegonial neck separate from one another to form an opening. Concurrently, the *neck and ventral canal cells* disintegrate to form an open passageway from the outside to the *egg.* Mature *sperm* are released by rupture of the *antheridium* and the upper layer of cells covering the gametophyte chamber in which the *antheridia* are enclosed. This occurs when the *gametophyte* is flooded with water, as during or immediately after a rain, since the motile *sperm* must have water in which to swim. *Sperm* are attracted to receptive *eggs* within mature *archegonia* by sex hormones released when the *neck and ventral canal cells* disintegrate.

Color the archegonium, the portion of the gametophyte, the zygote (H), and the embryo (I) on the two diagrams of archegonia at lower left. Also color the mature sporophytes (J) emerging from the gametophyte (B) and the foot (J¹), sporocytes (K), and meiospores on the two large diagrams at upper left.

Fusion of a *sperm* with an *egg* within the chamber of the *archegonium* produces a *zygote* that undergoes mitosis to form a multicellular *embryo* (sporophyte) contained within the *archegonium.* As with the other two groups of bryophytes, the mosses and liverworts, development of the *sporophyte* is completely dependent upon the *gametophyte.* In mosses and hornworts, the *sporophyte* is green and photosynthetic in its younger stages. Water and nutrients are absorbed from the *gametophyte* by the enlarged base, or *foot,* of the *sporophyte,* which functions as an absorbing organ.

As they grow, the columnar *sporophytes* emerge from the *archegonia* embedded within the upper surface of the *gametophyte.* The columnar *sporophytes* continue to elongate from their bases since new tissue is continually added by an area of active cell division, or intercalary meristem, near the base of each *sporophyte.* Within the outer protective wall of a *sporophyte* is a central column of sterile tissue called the columella (not separately colored). Surrounding the columella is an area of fertile tissue composed of diploid *sporocytes* and other diploid cells (not shown). The *sporocytes,* which are continually produced, undergo meiosis to form *meiospores* well above the intercalary meristem region within the maturing portion of the *sporophyte. Meiospore* formation is a continual process.

The *meiospores,* initially in clusters of four (tetrads), separate from one another as they mature. When the *meiospores* near the tip of a *sporophyte* are mature, the outer protective wall of the *sporophyte* splits into segments that separate from one another. As maturation of the *meiospores* within an elongating columnar *sporophyte* progresses, separation of the outer wall segments continues to keep pace with *meiospore* maturation. Some of the diploid cells within the spore-forming tissue that do not form *meiospores* function in the nutrition of the developing *sporocytes,* but others form unicellular and multicellular structures called pseudoelaters that may function in *meiospore* dispersal. As the *sporophyte* matures and releases *meiospores,* the pseudoelaters may rapidly uncoil to throw the *meiospores* from the *sporophyte.* The fine, dry, lightweight *meiospores,* which function as the primary dispersal unit, are then caught and carried by air currents.

ANTHOCEROUS LIFE HISTORY.

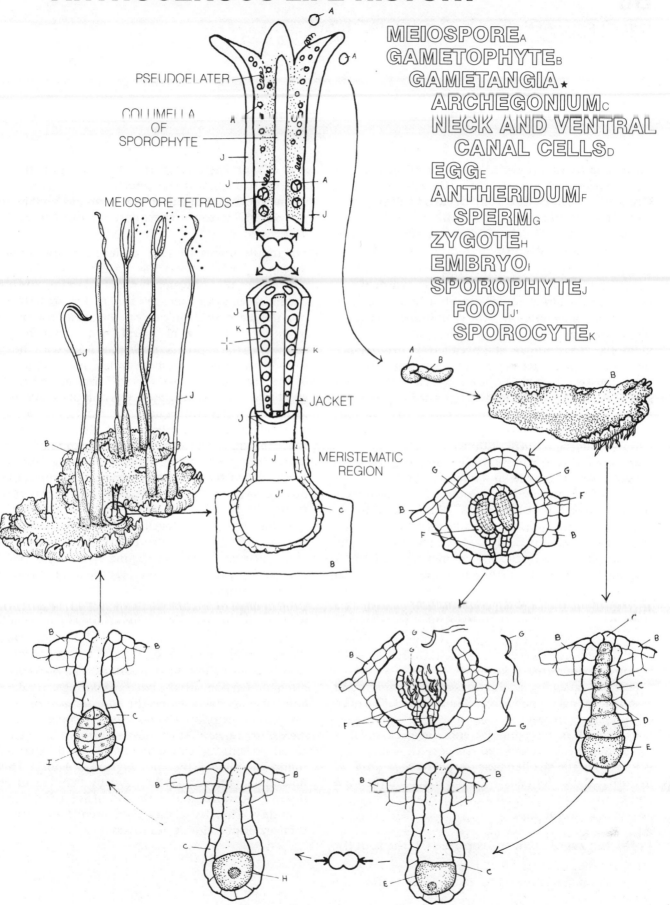

PSEUDOELATER

COLUMELLA
OF
SPOROPHYTE

MEIOSPORE TETRADS

JACKET

MERISTEMATIC
REGION

MEIOSPOREA
GAMETOPHYTEB
GAMETANGIA ★
ARCHEGONIUMC
NECK AND VENTRAL
CANAL CELLSD
EGGE
ANTHERIDUMF
SPERMG
ZYGOTEH
EMBRYOI
SPOROPHYTEJ
FOOTJ'
SPOROCYTEK

54
MOSSES (BRYOPHYTA)

Color the meiospores (A), protonema (B¹), leafy gametophyte (B²), antheridia (C), sperm (D), archegonia (E), neck and ventral canal cells (F), and egg (G) along the right side of the plate. Stop coloring when you finish with the archegonium immediately preceding the syngamy symbol.

The moss *Funaria,* class Musci, provides a good example of a typical life history for a moss, which like all land plants is diplohaplontic with a heteromorphic alternation of generations. The germination of a moss *meiospore* first produces a branched, uniseriate, filamentous *gametophyte* stage called a *protonema.* Buds of the *leafy gametophyte* stage soon develop along the *protonema.* The *gametophyte* generation is green and photosynthetic. The upright *leafy gametophytes* send uniseriate filaments called rhizoids (not separately colored) into the substrate for anchorage and for water and nutrient uptake.

Depending on the moss species, the *leafy gametophytes* are either male or female or both. In most mosses, the gametangia are produced within clusters of leaves at the tips of branches. Within the clusters, numerous sterile, uniseriate filaments called paraphyses (not colored separately) are scattered among the gametangia. In male clusters, numerous *antheridia* are usually grouped together at the end of a branch with diverging leaves that give a flowerlike appearance to the branch tip. An *antheridium* consists of a short stalk and a single layer of sterile jacket cells that surround the developing *sperm.* Numerous biflagellated *sperm* are released from each *antheridium* by rupture of the *antheridium* wall.

Female gametangia, the *archegonia,* are found among terminal branch leaves that are erect. The neck of immature *archegonia* is closed by neck cover cells (not shown separately) and the neck canal is plugged by *neck* and *ventral canal cells.* A single *egg* is located in the swollen base of the *archegonium.* At maturity, the cover cells open outward and the *neck* and *ventral canal cells* break down to create an open passageway to the *egg.*

Color the zygote (H), sporophyte (I), the foot (J), seta (K), and sporangium (L) of the sporophyte, and the calyptra (E¹), operculum (L¹), and peristome teeth (L²) of the sporangium, as well as the previously listed structures, on the remaining diagrams.

The motile *sperm* are released into the environment during a period of wetness, as from rainfall. They must have a film of water in which to swim. Through chemical sex attractants released by mature *archegonia,* the *sperm* find their way to the opening at the top of the *archegonium* neck and swim down the passageway to the *egg.* One *sperm* then fuses with the *egg* to form a *zygote.* The *zygote* undergoes mitosis within the archegonial chamber to form an embryonic *sporophyte.* The *sporophyte,* which is photosynthetic and partially dependent upon the *leafy gametophyte,* absorbs water and nutrients from the *leafy gametophyte* through an absorption structure called the *foot.* Above the *foot* region is an elongating, stalklike region called the *seta,* which carries the meiospore-producing *sporangium* well above the *leafy gametophyte.*

As the *sporophyte* develops within the *archegonium,* the *archegonium* also enlarges, but it does not keep pace with the developing *sporophyte.* Rapid elongation of the *seta* often tears the neck portion of the *archegonium* from the base, and it may remain as a cap of dead *gametophyte* tissue, called the *calyptra,* that covers the apex of the *sporangium* until maturity. The meiospore-forming portion of the *sporophyte,* the *sporangium* or capsule, has a caplike covering over its apex called the *operculum.* The *operculum,* often hidden beneath the *calyptra* during early development, falls away when the *sporangium* matures. The loss of the *operculum* opens the *sporangium* for *meiospore* release. Most moss *sporangia* have specialized structures, called *peristome teeth,* that surround the rim of the *sporangium* opening. These function in *meiospore* dispersal by bending in and flipping outward under the influence of changing humidity. The dry, lightweight *meiospores* are the primary dispersal unit of mosses.

FUNARIA LIFE HISTORY.

SEXUAL REPRODUCTION ★
MEIOSPORE A
GAMETOPHYTE B( )
   PROTONEMA STAGE B¹
   LEAFY GAMETOPHYTE B²
    ANTHERIDIUM C
     SPERM D
    ARCHEGONIUM E
      NECK AND VENTRAL
       CANAL CELLS F
      EGG G
ZYGOTE H
SPOROPHYTE I
  FOOT J
  SETA K
  SPORANGIUM L
   CALYPTRA E¹
   OPERCULUM L¹
   PERISTOME TEETH L²

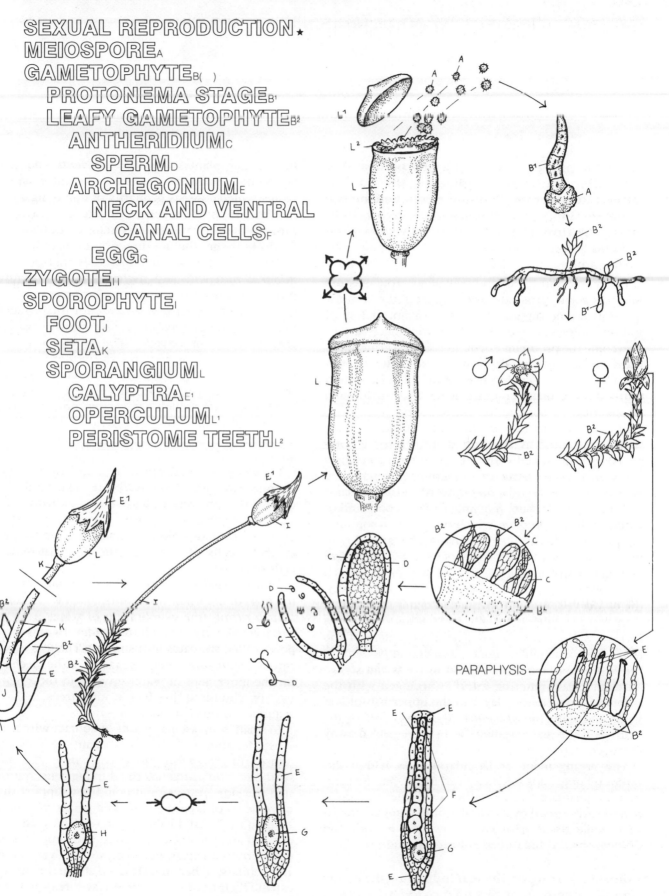

PARAPHYSIS

Color the pith (A), xylem (B), and phloem (D). Also color the arrows indicating the flow of water and minerals (C) into the root and, on the "stem section," up the stem and the arrows indicating the flow of food materials (E) down the leaf and down the stem section. Do not color the cortex and epidermis (F).

In order for land plants to break away from existing solely as small, prostrate plants appressed to a moist substrate, numerous adaptations, both morphological and physiological, evolved. These were necessary to overcome the problems encountered with growth upward into a relatively dry atmosphere without external support. The development of the vascular system fulfilled two major requirements by providing a means for both conduction and support.

In a typical dicot stem, the central portion of the stem is occupied by the *pith,* a storage tissue composed of thin-walled cells. Surrounding the *pith* is a cylinder of *xylem* tissue that functions in conducting *mineral nutrients* and *water* upward and as the major structural plant tissue. A cylinder of a second kind of vascular tissue, *phloem,* which functions in conducting photosynthetic *food materials,* primarily downward, and as a support tissue, surrounds the *xylem* cylinder. Surrounding all this is the *cortex,* which functions in storage and support, and an outer layer, the *epidermis,* which functions in water retention.

In roots, a *pith* is usually lacking and the *xylem* forms a solid central core. The arrangement of the two types of vascular tissues, *xylem* and *phloem,* in the leaf differs from that found in roots and stems. Vascular tissue entering a leaf is arranged with the *xylem* as a continuous layer in the upper half of leaf veins and with the *phloem* in the lower half of the veins as if the leaf vascular tissues were peeled away from the stem.

The arrangement of vascular tissues within different plant groups has many variations, but the commonest arrangements are either a cylinder or core of *xylem* surrounded by a cylinder of *phloem* or strands of vascular tissue with *xylem* toward the inside and *phloem* toward the outside of each strand.

Color the arrows on the leaf indicating the direction of movement of carbon dioxide (G), oxygen (H), and transpiration water (I).

In vascular plants, *mineral nutrients* and *water,* taken into young roots near their tips, enter the *xylem* tissue, which consists primarily of nonliving conductive and supportive cells, and are carried upward through the root and stem systems to all plant parts. In *xylem* tissue, the net direction of material flow is upward, and the primary materials transported are *mineral nutrients* and *water.* The photosynthetic process that occurs in all green portions of a plant requires *water* and *carbon dioxide* for the production of photosynthetic *food materials,* mostly sugars, and the liberation of *oxygen.* The synthesis of many organic compounds requires the *mineral nutrients* carried in the *water.* All living cells require an energy source, and this is provided by the photosynthetic *food materials,* which are carried downward from the elevated sites of photosynthesis through the *phloem* tissue to nonphotosynthetic cells in the stem and roots.

In *phloem* tissue, the primary materials transported are photosynthetic *food materials.* The net direction of flow is from areas of high *food materials* concentration, or sources, to areas of low *food materials* concentration, or sinks (for example, from the photosynthetic cells of leaves to the nonphotosynthetic cells of roots).

Photosynthesis requires a large amount of water, but most of the water taken into a plant is lost through *transpiration* (the evaporation of water from plant surfaces) directly through the *epidermis* and through pores called stomates (not shown). The loss of water through *transpiration* is one factor of the mechanism for the movement of water upward in the *xylem,* to over two hundred fifty feet in some trees, and also is required for normal plant growth.

In small, nonvascular plants in contact with a moist substrate, simple diffusion is a sufficient means of transporting *mineral nutrients, water,* and photosynthetic *food materials* to all living cells, but diffusion is a slow process and could not support the requirements of a large plant. The plants that are best adapted to the land habitat, the vascular plants, have evolved an efficient conduction and support system that permits them to be successful in even the most arid habitats, to become elevated above the surface to expand their photosynthetic surface area and to reach for the light.

GENERAL VASCULAR SYSTEM.

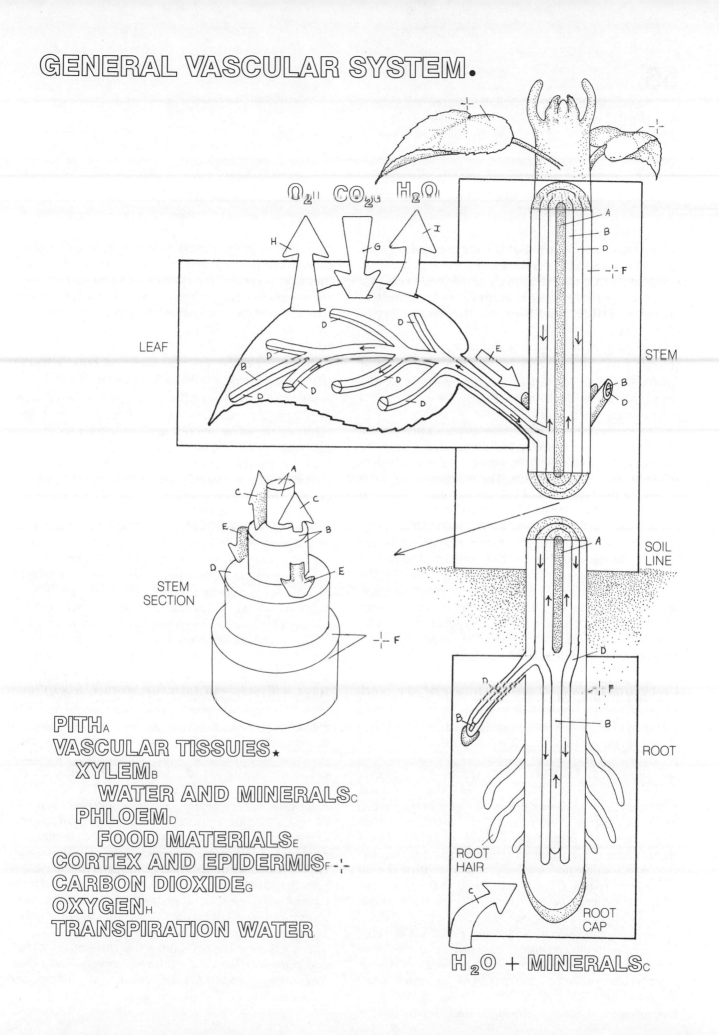

O_2 CO_2 H_2O

LEAF

STEM

STEM SECTION

SOIL LINE

ROOT

ROOT HAIR

ROOT CAP

PITH_A

PITH_A
VASCULAR TISSUES★
   XYLEM_B
      WATER AND MINERALS_C
   PHLOEM_D
      FOOD MATERIALS_E
CORTEX AND EPIDERMIS_F-¦-
CARBON DIOXIDE_G
OXYGEN_H
TRANSPIRATION WATER

H_2O + MINERALS_C

Color the epidermis (A) of the stem cross section and (B) through (D) on the boxed enlargement of epidermis (A), and the single epidermis cell (A). The small squares on the stem cross section illustrate possible positions of the tissue types discussed.

Simple tissues are composed of a single cell type. One simple tissue, the *epidermis,* which is composed of *epidermal cells,* covers the entire surface of the primary plant body. In most plants, the *epidermis* is only one cell layer thick.

A major function of the *epidermis* is to protect the plant from excessive water loss due to desiccation. To do this, *epidermal cells* have a *primary cell wall* that is thickest on the external side. The *primary cell wall* is started at the beginning of cell growth and completed soon after the cell attains its mature size and shape. *Protoplasts* of *epidermal cells* secrete a waxy substance, called cutin, that forms a layer, called the *cuticle,* on the surface of the *epidermis.* The cuticle forms a barrier to water movement through the *epidermis.* Epidermal cell shape is diverse, but many appear like pieces of a jigsaw puzzle in surface view. *Epidermal cells* are living at maturity and have a functional *protoplast,* but they are typically non-photosynthetic and lack chloroplasts.

Color the cortex (E) and pith (E¹) of the stem cross section, the boxed enlargement of parenchyma (F), and the single parenchyma cell (F).

Parenchyma cells have a thin *primary cell wall* and a functional *protoplast* at maturity. *Parenchyma cells* are present as individual cells or as *parenchyma tissue* in almost all plant parts, but they are most prevalent in the *cortex* and *pith.* The *primary cell wall* of *parenchyma cells* is uniformly thin throughout. A diagnostic feature of *parenchyma tissue* is the presence of air spaces of various sizes between adjoining *parenchyma cells* that form a continuous network for gas exchange throughout the plant. In general, *parenchyma cells* are many-sided, roughly spherical and come in a broad range of sizes.

Parenchyma functions in water and food material storage, limited intercellular transport, and photosynthesis when chloroplasts are present. Through hydrostatic pressure, *parenchyma* provides some support for soft plant structures. A drop in hydrostatic pressure, through insufficient water,

results in wilting. *Parenchyma cells* in green plant parts contain numerous chloroplasts for photosynthesis. This specialized type of *parenchyma* is called chlorenchyma (not shown), which means "green tissue." Most *parenchyma cells* retain their capacity to divide.

Color the boxed enlargement of collenchyma (H) and the single collenchyma cell (H).

Some cell types, especially those that function in support, produce materials that are added in layers to the inside of the *primary cell wall* to form a secondary cell wall. *Collenchyma* cells form a secondary cell wall consisting primarily of *pectin* (pectic compounds). The secondary cell walls of most *collenchyma cells* are unevenly thickened, with thicker areas occurring in the corners. Others have evenly thickened secondary cell walls. The primary function of *collenchyma* is the support of herbaceous plant parts. *Collenchyma* is typically found in the periphery of the stem *cortex* and in leaves. At maturity, *collenchyma cells* contain a functional *protoplast* that may contain chloroplasts. *Collenchyma cells* are usually elongate in the direction of the stem axis and have tapered, oblique ends.

Color the two types of sclerenchyma (J): the fiber and sclereid (asterosclereid). Also color the vascular tissues (M) on the stem cross section.

A fourth simple tissue, called *sclerenchyma,* is composed of *sclerenchyma cells,* which have secondary cell walls consisting of cellulose and *lignin,* an extremely hard and durable substance. The secondary cell walls are typically evenly thickened. *Sclerenchyma* is a major support tissue. Two types of *sclerenchyma cells,* fibers and sclereids, are commonly recognized. Fibers are extremely long cells that usually form bundles of *sclerenchyma.* Sclereids are short cells that come in a variety of shapes. In most *sclerenchyma cells,* there is no *protoplast* at maturity, only an empty space, called the *lumen,* that the *protoplast* occupied before it died and was reabsorbed.

The vascular region, unlike the *epidermis, cortex,* and *pith,* consists of a mixture of several cell types. Thus, the *vascular tissues,* xylem and phloem, are considered to be complex tissues.

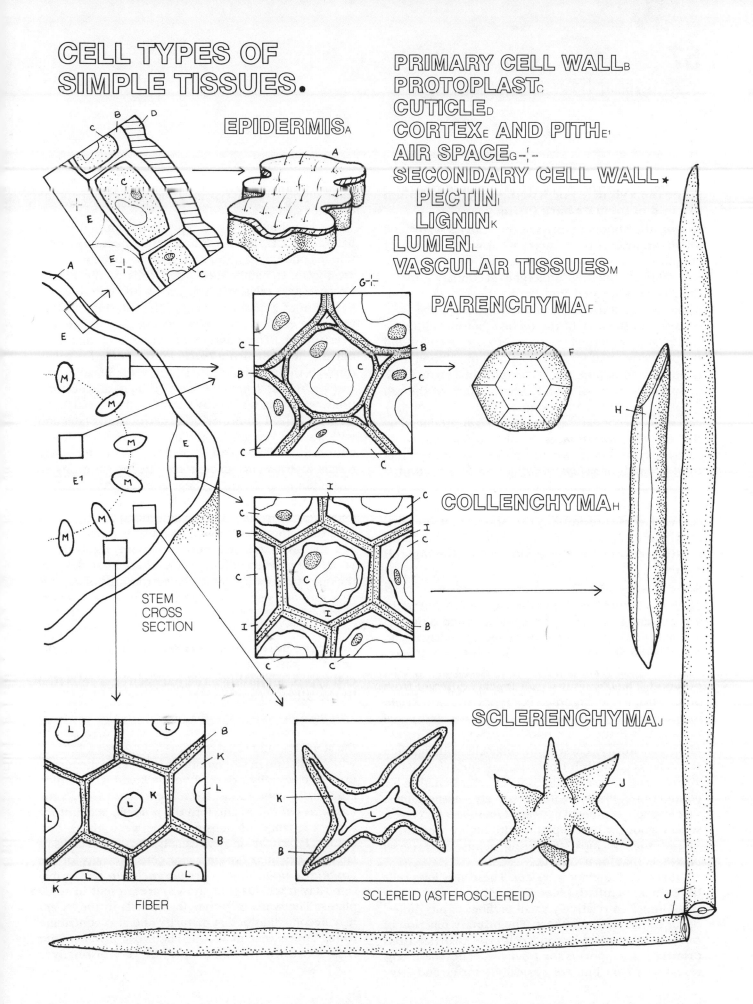

CELL TYPES OF SIMPLE TISSUES.

PRIMARY CELL WALL_B
PROTOPLAST_C
CUTICLE_D
CORTEX_E AND PITH_E'
AIR SPACE_G-¦-
SECONDARY CELL WALL ★
PECTIN_I
LIGNIN_K
LUMEN_L
VASCULAR TISSUES_M

EPIDERMIS_A

PARENCHYMA_F

COLLENCHYMA_H

SCLERENCHYMA_J

STEM
CROSS
SECTION

FIBER

SCLEREID (ASTEROSCLEREID)

COMPLEX TISSUES

On the diagram of a wedge from a woody stem, color the xylem (A) or wood, vascular cambium (B), and phloem or bark (using gray).

Xylem, the conductive tissue that carries water and mineral nutrients from the roots upward, is a highly complex tissue consisting of several cell types. In a woody stem, one that undergoes secondary growth, the *xylem* is located internal to the *vascular cambium.* External to the *vascular cambium* is the bark, consisting primarily of the *phloem* but including the cortex and epidermis on younger stem portions and the periderm on older stem portions. In addition to its conductive function, the *xylem* is the main structural, or support, tissue of woody plants. Many of the cell types in *xylem* have thickened, lignified, secondary cell walls. Secondary *xylem,* the *xylem* produced by the *vascular cambium,* is the plant tissue commonly referred to as wood. We take advantage of the structural strength of *xylem* by using it in construction.

Color the fiber (C), fiber tracheid (D), tracheid (E), and vessel members (F), starting at the bottom right corner of the plate. Choose contrasting colors for these and the following cell types.

Several types of cells, forming a gradient of morphological types having lignified secondary cell walls, are present in *xylem.* The cellular composition of *xylem,* with regard to the presence, abundance, and distribution of each cell type, contributes significantly to the characteristic appearance of a particular type of wood. Thus, walnut wood appears different from pine or oak wood. Lignified cell types found in *xylem* include *fibers, fiber tracheids, tracheids,* and *vessel members.* All these are nonliving, without a protoplast, at maturity. These nonliving cells with lignified secondary cell walls form the bulk of the *xylem* tissue. *Vessel members* are the epitome of cell adaptation for conduction. Plants with relatively unspecialized *xylem,* such as pines, lack *vessel members* (have vesselless wood), whereas plants with highly specialized *xylem,* such as most flowering plants, have *vessel members* that form vessels.

Fibers are abundant in *xylem.* These very long cells have thick, lignified, secondary cell walls and a very small lumen. A relatively small number of pits (holes) occur in the walls of *fibers.* The primary function of *fibers* is support. A second cell type that functions primarily in support is the *fiber tracheid.* These cells resemble *fibers* but are typically shorter and have thinner lignified secondary cell walls. One or both of these supportive cell types may be present, depending on plant species.

The remaining two cell types with lignified secondary cell walls, *tracheids* and *vessel members,* function in conducting water and mineral nutrients. These are the most highly specialized cells of the *xylem.* *Tracheids* are relatively long cells with numerous pits but without a distinct opening (perforation) at each end. The ends of *tracheids* are usually tapered and overlapping. The passage of water and dissolved mineral nutrients from *tracheid* to *tracheid* is solely through the matching pits in the walls of adjoining *tracheids.* *Vessel members* have both pits and perforations, which are divided by bars of wall material in the less-specialized, multiperforate perforation plates and with a single opening in the more specialized simple perforation plates. Less-specialized *vessel members* resemble *tracheids* except for the perforations in their oblique ends. Highly specialized *vessel members* are short and broad with little or no angle to their ends. *Vessel members* are stacked end to end to form a long tube, the vessel, much like a soda straw. Water and dissolved mineral nutrients move freely from *vessel member* to adjoining *vessel member* through the open perforation plates. Numerous pits of various structure are also present in the side walls of *vessel members.*

Color the parenchyma ray (G) at the bottom left and the parenchyma rays in the wedge and all cell types, including axial parenchyma cells (G¹) in the block of xylem tissue.

Parenchyma cells are also present in the *xylem* as stacks of *axial parenchyma* cells in the direction of the stem axis and as lateral *parenchma rays,* both of which function in storage and in intercellular conduction. The laterally oriented *parenchyma rays,* which are usually continuous, from the *xylem* through the *vascular cambium* and into the *phloem,* account for some exchange of substances between these two tissues. *Parenchyma rays* usually consist of stacks of laterally oriented *parenchyma* cells. Activity of the *vascular cambium* initiates *parenchyma ray* formation. Ray tracheids (not shown) are present in some plants. The walls of *parenchyma* cells in the *xylem* may become lignified at maturity and also contribute to the appearance of wood. The broad *parenchyma rays* of oak wood, for example, are very distinctive.

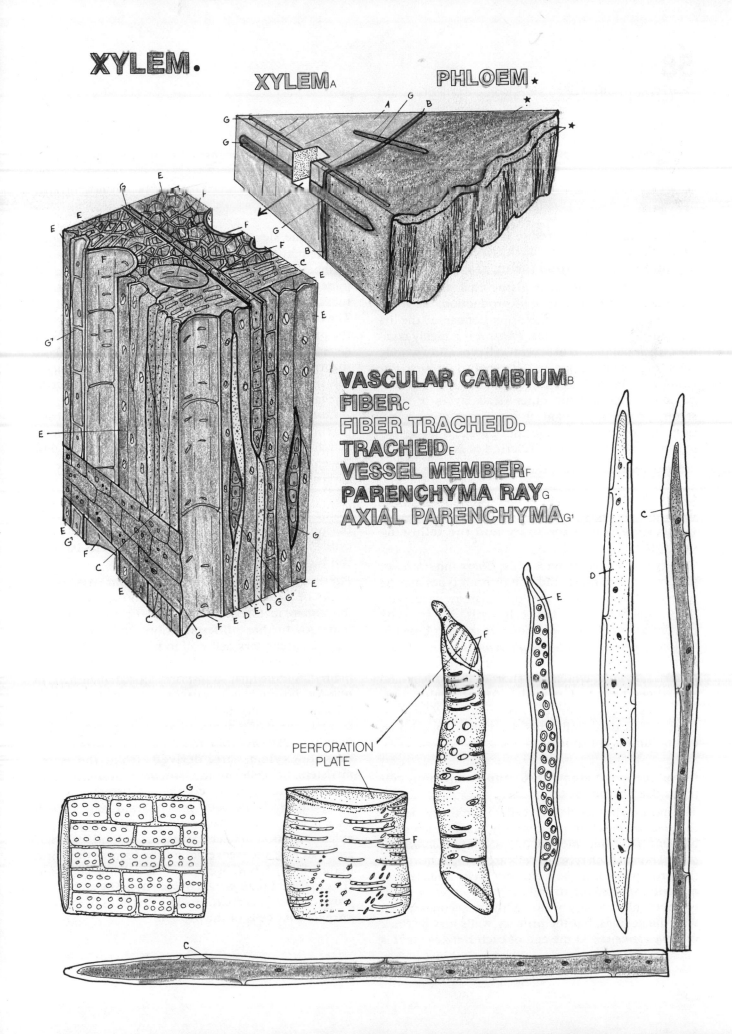

XYLEM.

XYLEM A   PHLOEM ★

VASCULAR CAMBIUM B
FIBER C
FIBER TRACHEID D
TRACHEID E
VESSEL MEMBER F
PARENCHYMA RAY G
AXIAL PARENCHYMA G'

PERFORATION
PLATE

58
COMPLEX TISSUES

On the diagram of a wedge from a woody stem, color the phloem (A) or bark, vascular cambium (B), and xylem or wood (using gray).

Phloem is the conductive tissue that carries food materials from their source of production, the green photosynthetic areas, to their point of use, all the living cells, or to storage sites. *Phloem* is a highly complex tissue composed of several cell types. In a woody stem, the *phloem* is located external to the *vascular cambium*. In addition to its conductive function, *phloem,* especially the outer dead layers of older stems, provides mechanical and water loss protection. In most instances, the entire *phloem* tissue, living and dead, is commonly referred to as the bark of a plant.

Color the fiber (C) and sclereid (D), starting at the bottom right corner of the plate. Choose contrasting colors for these and the following cell types.

Two types of sclerenchyma cells, *fibers* and *sclereids,* may occur in *phloem,* and one or both types may be present. *Fibers,* which are long, narrow cells with thick, lignified, secondary cell walls, are typically present as bundles of numerous *fibers* distributed in various patterns within the *phloem*. In some plants, these *fibers* provide a commercial source of textile *fibers*. *Sclereids,* which are shorter than *fibers* and sometimes branched, are also found, and a morphological gradient between *fibers* and *sclereids* often exists. Both these cell types are nonliving, with an empty lumen, at maturity.

Color the sieve element (E) and companion cell (F) at the bottom of the plate.

The most highly specialized cell within the *phloem* is the *sieve element*. In most woody plants, each *sieve element* is accompanied by one or more *companion cells*. Though both types of cells are living at maturity, the protoplast of a *sieve element* lacks a nucleus at maturity. *Sieve elements* and *companion cells* typically have only primary walls, with no secondary cell wall thickenings, but the primary walls may be thick in *sieve elements*. At the end of each *sieve element* is

an oblique sieve plate with numerous perforations, which may be oblong or circular. Like the vessel members of the xylem, the *sieve elements* are adjoined end to end. Stacks of *sieve elements* form sieve tubes through which food materials are transported. The protoplast of one *sieve element* is connected with the protoplasts of the adjoining *sieve elements*. *Sieve elements* are continuous by strands of protoplasm that connect the protoplasts through the openings in the sieve plates. The side walls of *sieve elements* typically have pits.

Companion cells are thin walled, specialized, parenchymalike cells that accompany *sieve elements* and are interdependent with them. However, some lower vascular plants lack *companion cells*. Death of a *sieve element* results in death of the *companion cell*.

Color the parenchyma ray (G) at the bottom left, the parenchyma rays in the wedge, and all cell types, including axial parenchyma cells (G¹), in the block of phloem tissue.

Parenchyma cells occur in *parenchyma rays* and as stacks of *axial parenchyma* cells in the *phloem*. As the *phloem* matures, the *axial parenchyma* cells may undergo further differentiation and produce a lignified secondary cell wall to form *sclereids*. In the *phloem, parenchyma* cells function in storage and limited intercellular transport. Stored materials in *phloem parenchyma* cells include starch, various substances in crystalline form, and tannins, the natural source of tannic acid for leather preservation (tanning). *Parenchyma rays,* like the *parenchyma rays* in xylem, are derived from the same meristematic cells in the *vascular cambium* and therefore are typically continuous between xylem and *phloem*. In *phloem, parenchyma rays,* which have a structure similar to the *parenchyma rays* in xylem, function in lateral transport of food materials and water. As the stem enlarges, the fissures that form in the older *phloem* become filled with *parenchyma* cells to form broad *parenchyma rays*. The additional *parenchyma* cells are formed by divisions of existing *parenchyma* cells of the *phloem*.

PHLOEM.

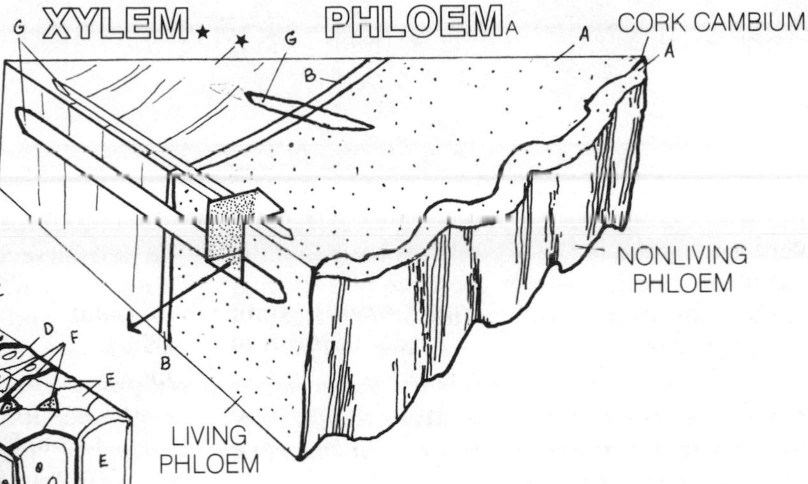

XYLEM ★   PHLOEM A   CORK CAMBIUM

NONLIVING PHLOEM

LIVING PHLOEM

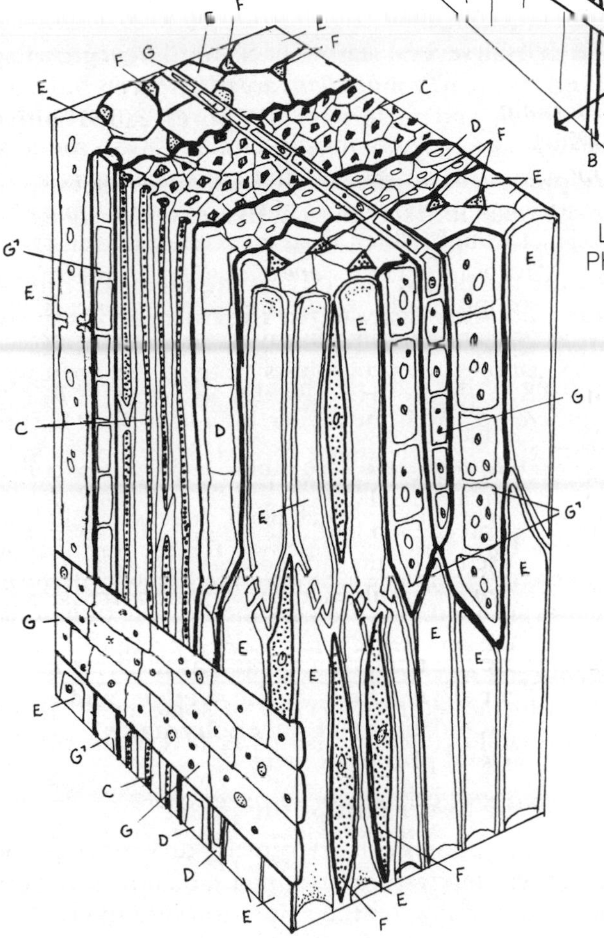

VASCULAR CAMBIUM B
FIBER C
SCLEREID D
SIEVE TUBE ★
  SIEVE ELEMENT E
  COMPANION CELL F
PARENCHYMA RAY G
AXIAL PARENCHYMA G'

SIEVE PLATE

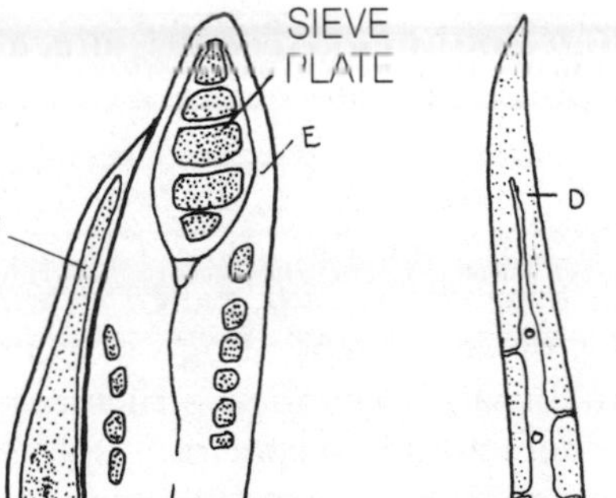

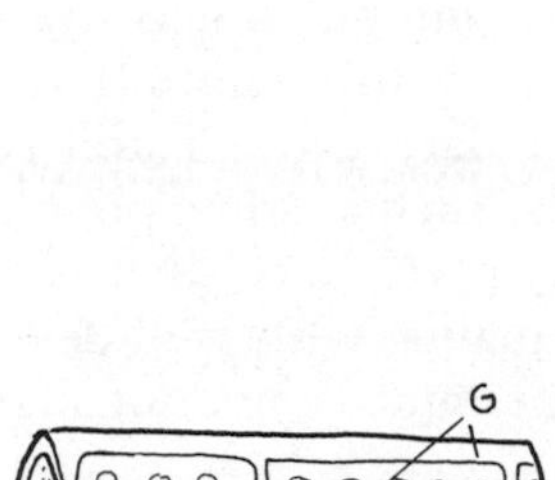

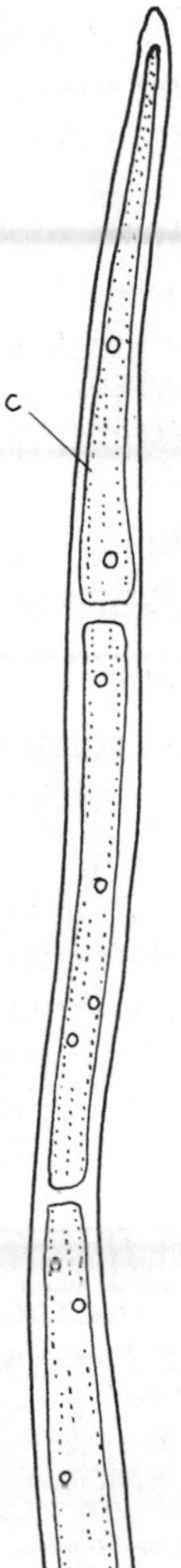

VASCULAR PLANT STRUCTURE

The plant groups above the bryophytes are called the vascular plants because of the presence of a specialized conduction system for the transport of water and food materials from source to site of use.

Color over the root (A) on the plant diagram at the upper left. On the enlargement of the root tip, color the root (A) and root cap (B).

Major *root* functions include water and mineral uptake, conduction, and support of the shoot by anchoring the plant to the substrate. The *root cap* at the tip of each *root* protects the tender root tip as it grows. In most vascular plants, the root system is below the substrate surface and nonphotosynthetic.

Color the stem (C) and leaves (D) on the plant diagram and the nodes (E), internodes (F), petioles (G), lamina (H), stipules (I), axillary buds (J), leaf scars (K), and lenticels (L) on the enlarged diagrams to the right.

The shoot system consists of a branched or, less commonly, an unbranched *stem* that functions in conduction and in the support and elevation of the photosynthetic structures, *stems* or *leaves* or both, for optimal light reception. Elevation of a plant above the substrate by upright *stem* growth allows for more photosynthetic surface due to vertical stacking of photosynthetic structures. *Stems* may be divided into two morphological regions: *nodes* and *internodes*. A *node* is a stem region where one or more *leaves* is attached. An *internode* is the portion of *stem* between two adjacent *nodes*.

Most *leaves* have a stalklike region, the *petiole*, that supports an expanded region, the *lamina* or blade. Outgrowths on the base of the *leaf*, called *stipules*, may or may not be present, depending on species. *Stipules* vary from broad, expanded structures to sharp, rigid spines, depending on species.

Immediately above each *leaf,* at the juncture of the *stem* and *leaf,* are one or more *axillary buds*, which are dormant, meristematic primordia (buds) that may develop into a vegetative shoot, flowering branch, adventitious root, or other structure. Each lateral *stem,* or branch, of most plants began development as an *axillary bud* that became active.

In vascular plants that shed their *leaves* each season, known as deciduous plants, a scar, called a *leaf scar,* covered with a corky protective layer, marks the position of the fallen *leaves. Leaf scars* are visible,

often from several seasons back, until *stem* growth obliterates them. Young *stems* are often green and photosynthetic, and numerous small pores, called *lenticels,* which permit gas exchange for photosynthesis and respiration, are present in *internode* regions. *Lenticels* open into the internal tissues of the *stem.*

Color the bud scales (M) on the terminal bud and the bud scale scars (N) at the bottom of the stem diagrams.

During active growth, the tips of the *stem* produce new *stem* growth and *leaves;* but as the plant becomes dormant, the *internodes* at each *stem* tip no longer elongate, and a dormant terminal bud is formed. Closely clustered, specialized leaves, called *bud scales,* form a protective cover for the terminal bud. When the *stem* once again becomes active, all the *bud scales* are shed, leaving a distinctive ring of *bud scale scars*. One set of *bud scale scars* is produced by each *stem* each season.

An introductory overview of some vascular plant forms is illustrated on the lower half of the plate.

Color the lower half of the plate.

Stem morphology is extremely diverse, but most plants can be classified as annuals, biennials, or perennials, depending on their duration (life span), and as herbs, vines (not shown), shrubs, or trees, based on *stem* woodiness and growth form. Annuals live for one season (less than one year) only; biennials begin growth in one year and complete growth and then die the following year; and perennials live for a number of years (a few to a few thousand). Herbs lack or have very limited secondary growth and relatively soft stems. All annuals, such as some poppies, are herbaceous, as are many perennials, such as most lilies and ferns. All woody plants, with well-developed secondary growth, are perennials, but not all perennials are woody.

Shrubs are woody plants with a shoot system that branches into few to numerous main *stems* at or near the base, and they are usually low growing. Trees are woody plants with a shoot system that has a single main *stem,* or trunk, that may remain unbranched except for smaller lateral branches, as in most pines, or become highly branched, as in maples. Lateral branches produced on the lower trunk usually die back, due to shading by upper branches, and are shed from the tree.

VEGETATIVE ORGANOGRAPHY.

ROOT_A
ROOT CAP_B
SHOOT ★
STEM_C
   NODE_E
     INTERNODE_F
LEAF_D
   PETIOLE_G
   LAMINA_H
   STIPULE_I
AXILLARY BUD_J
LEAF SCAR_K
LENTICEL_L
TERMINAL BUD ★
   BUD SCALE_M
   BUD SCALE SCAR_N

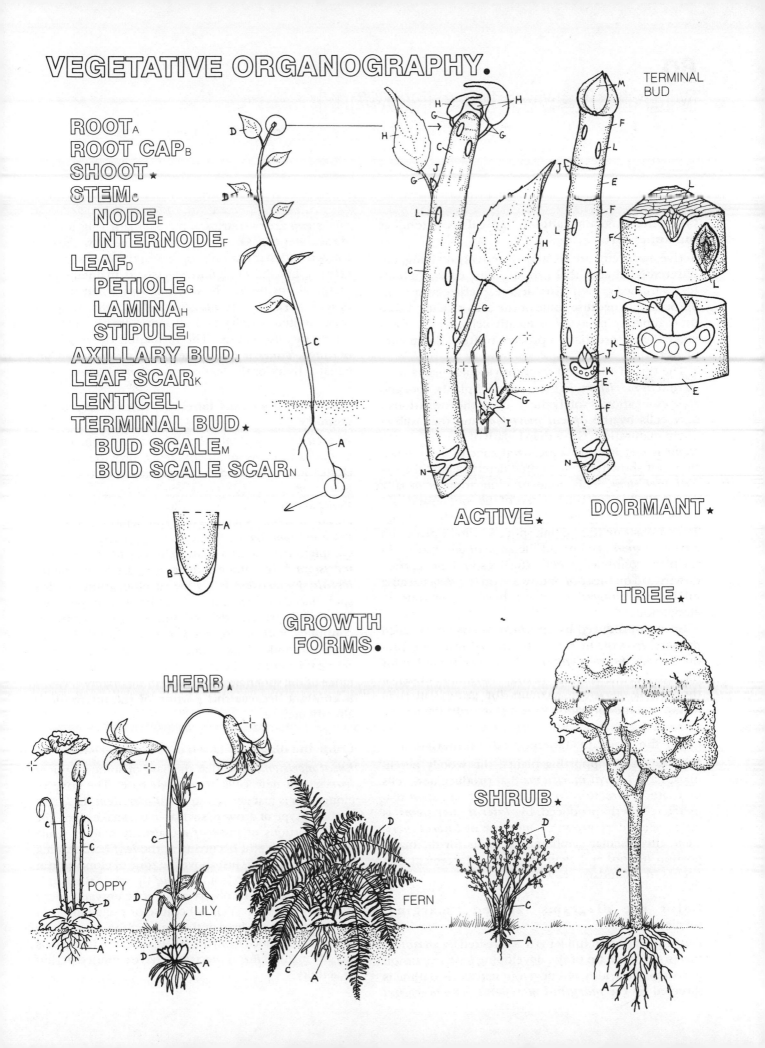

TERMINAL BUD

ACTIVE ★

DORMANT ★

GROWTH FORMS.

HERB ★

POPPY

LILY

FERN

SHRUB ★

TREE ★

Color the diagrams labeled "apical meristem" at the upper left.

Active meristems are clusters of actively dividing, undifferentiated cells that produce new cells as a basis for continued growth. To a degree, meristems are self-perpetuating because some of the cells produced continue functioning as meristematic cells, while others enter the differentiation process to form the mature plant tissues.

The tips of *roots* and *stems* of vascular plants have an *apical meristem*. In most vascular plants, *root* and *stem* elongation is solely the result of the formation of new cells by the *apical meristem* and their subsequent maturation. Growth in length of most *roots* and *stems* is restricted to a localized zone of cell elongation (not shown) located immediately behind the *apical meristem*. This zone of elongation is usually within the last few centimeters of the root tip and usually within the last several centimeters of the stem tip. To demonstrate this, an ink spot (●) can be placed on a *root* or *stem* just behind the zone of elongation. As the plant continues growth, the ink spot is not carried downward on the *root* or upward on the *stem* because no further elongation occurs beyond the zone of elongation.

Growth produced by an *apical meristem* is called primary growth. In *roots,* the *apical meristem* produces both the *root cap* and the cells responsible for the initial root diameter and root elongation. In *stems,* the *apical meristem* accounts for growth in stem length and the formation of leaf primordia (buds).

Color the diagrams labeled "lateral meristem."

Most tall, self-supporting plants, the woody perennials, have *lateral meristems* that produce new cells in a lateral direction to increase *root* and *stem* diameter. Growth produced by *lateral meristems* is called secondary growth. One type of *lateral meristem,* the vascular cambium, accounts for the major portion of *root* and *stem* diameter enlargement that occurs in secondary growth.

Color the diagrams labeled "marginal meristem."

Leaf elongation to full length is directed by an *apical meristem* at the tip of the developing *leaf;* expansion to its mature width and development of its outline is directed by the *marginal meristems*. The *marginal meristems* are two strips of meristematic cells located at the outer edges of the *leaf* margins. As new cells are formed and maturation progresses, the *leaf* width increases, but the *marginal meristems* maintain their marginal positions. The *leaf* matures when the leaf *apical meristem* and *marginal meristems* complete their limited activity and all meristematic cells differentiate and mature. The formation of irregularly outlined *leaves* is due to genetically controlled unequal activity of the leaf *marginal meristems*.

Color the diagrams labeled "intercalary meristem."

Intercalary meristems, located just above the younger nodes in many monocot *stems,* provide for *stem* elongation in addition to that produced by the *apical meristems*. The activity of several *intercalary meristems* accounts, in part, for the rapid growth of some monocot *stems,* such as corn and bamboo. *Intercalary meristems* have a limited period of activity. At the completion of activity, all cells of the *intercalary meristem* differentiate and mature. Just above the *intercalary meristem* is a zone of elongation. An ink spot placed on a *stem* in the mature part of the internode while growth, due to an *intercalary meristem,* continues, will be observed to move upward, away from the node. This is due to new growth formed above the node by the *intercalary meristem*. The distance of the spot below the node immediately above it is constant because this portion of the internode is already mature.

Color the diagrams labeled "basal meristem."

Basal meristems, found at the base of some monocot *leaves,* add new cells at the *leaf* base. The zone of elongation is just above the *basal meristem*. In *leaves* with this type of growth, such as in the snake plant, the upper portions of the *leaf* are already mature while new growth is still occurring at the *leaf* base. An ink dot placed on a *leaf* just above the zone of elongation at a particular distance from the *leaf* tip while the *leaf* is still growing will be observed to move upward, away from the *leaf* base. This is because of the addition of new growth at the base of the *leaf*. Yet the dot remains a constant distance from the *leaf* tip because that area is already mature. *Basal meristems* have a limited period of activity.

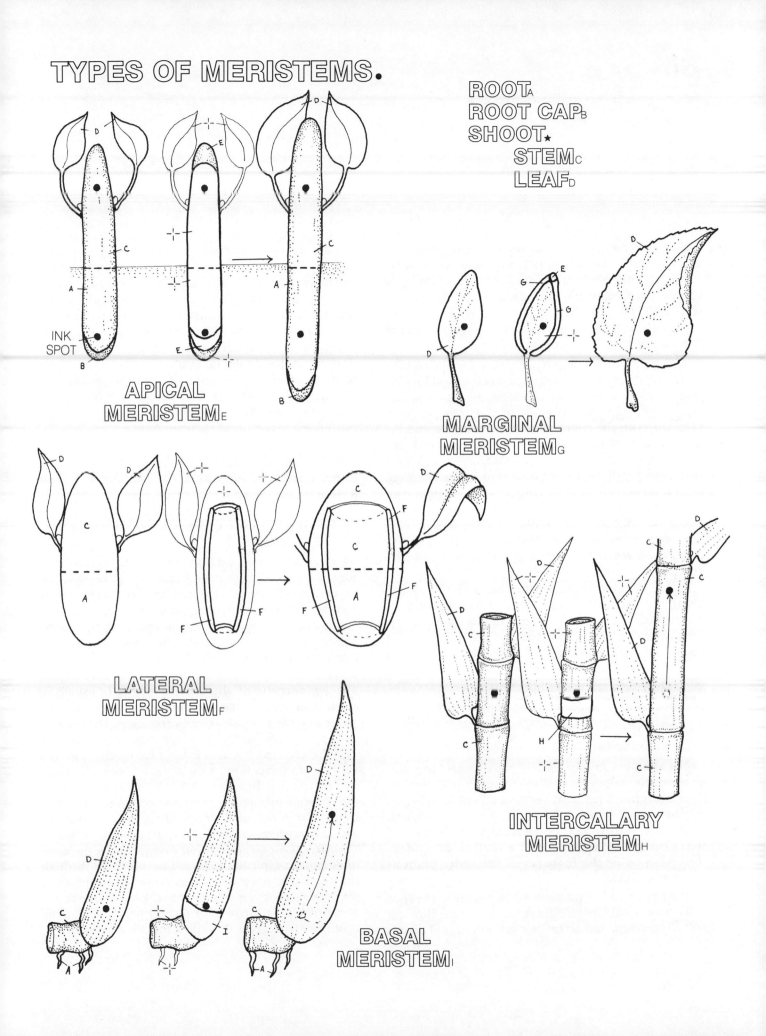

TYPES OF MERISTEMS.

ROOT_A
ROOT CAP_B
SHOOT★
STEM_C
LEAF_D

INK SPOT

APICAL MERISTEM_E

MARGINAL MERISTEM_G

LATERAL MERISTEM_F

INTERCALARY MERISTEM_H

BASAL MERISTEM_I

Color the root tip (A), zone of elongation (C), zone of differentiation (D), and mature root (F) on the young plant at the upper left and the root cap (A^1), apical meristem (B), zone of elongation (C), and the diagrammatic sequence of cell elongation (C^1) on the first two enlargements at the left of the root growth sequence.

In a root, growth in length occurs only in the vicinity of the *root tip*. A *root cap*, a protective and lubricative structure, covers, like a thimble, the tender *apical meristem* of the root. The *apical meristem* produces cells for both the *root cap* and the root. Behind the *apical meristem* is a region of pronounced *cell elongation*, called the *zone of elongation*, in which root cells produced by the *apical meristem* undergo enlargement, primarily in length. All growth in root length occurs in the *zone of elongation*. No root elongation occurs in the older root portions beyond the *zone of elongation*. The cells in the *zone of elongation* remain immature and relatively undifferentiated.

As the *root tip* is pushed through the soil by *cell elongation* occurring in the *zone of elongation*, cells on the outer surface of the *root cap* are worn away. This provides lubrication to aid in the movement of the *root tip* through the soil. New *root cap* cells, to replenish those lost during root growth, are continuously produced at the tip of the *apical meristem*. The *apical meristem* also continuously produces new meristematic cells and cells that will enter the maturation process in the *zone of elongation*.

Color the previously listed features and the diagrammatic representation of cell maturation (D^1), root hairs (E), and section of epidermis (E^1) on the last diagram of a root in the root growth sequence. Color the epidermal cells and root hairs on the enlargement of a section of epidermis. Also color the features of the older plant at the upper left.

At the upper end of the *zone of elongation*, the root cells reach their mature (maximum) length through *cell elongation*, and no further growth in root length

occurs. Beyond the *zone of elongation* is the root portion, called the *zone of differentiation*, in which *cell maturation* is completed. The cells of the various root regions, epidermis, cortex, and vascular cylinder, all become differentiated into their constituent cell types. In the *zone of differentiation*, the *epidermal cells* begin formation of *root hairs*, which function in water and mineral nutrient uptake. *Root hairs* form as narrow outgrowths of the outer wall of individual *epidermal cells*. During *root hair* formation, the epidermal cell nucleus (not separately colored) is positioned at the tip of the *root hair*.

Color the epidermis (G), cortex parenchyma (H), endodermis (I), pericycle (J), phloem (K), and xylem (L) on the full root cross section. Also color the previously listed structures and the lateral root (M) on the quarter sections at the bottom of the plate.

As viewed in cross section, the major regions of a root, from outside to center, are the *epidermis;* the cortex, consisting of an outer area of *parenchyma* and an inner *endodermis;* and the vascular core (cylinder), consisting of the *pericycle*, the *phloem*, and the *xylem*. The *epidermis* functions as a protective covering for the root. The *parenchyma* of the cortex functions in storing water and food materials. The *endodermis* of the cortex, which is usually a single cell layer thick, mediates the movement of water into the vascular core. Surrounding the *phloem* and *xylem* is the *pericycle*, which is the portion of the vascular core from which *lateral roots* arise. *Lateral roots* begin formation in the *pericycle* near the juncture of a *phloem* bundle and the tip of a *xylem* lobe. In *lateral root* formation, *pericycle* cells become meristematic to form the *root tip*, including the *root cap* and *apical meristem*, which begins growth outward. The *lateral root*, through *cell elongation* in the *zone of elongation*, pushes outward into the soil through the original root's tissues. This causes a disruption of the original root's tissues and a break in the protective epidermal layer. Growth of *lateral roots* follows the root growth process described for the main root.

PRIMARY ROOT GROWTH.

ROOT TIP A
ROOT CAP A¹
APICAL MERISTEM B
ZONE OF ELONGATION C
CELL ELONGATION C¹
ZONE OF DIFFERENTIATION D
CELL MATURATION D¹
ROOT HAIR E
EPIDERMAL CELL E¹
MATURE ROOT F

OLDER PLANT

YOUNG PLANT

ROOT GROWTH SEQUENCE

CELL NUCLEUS

EPIDERMIS G
CORTEX ★
PARENCHYMA H
ENDODERMIS I
VASCULAR CORE ★
PERICYCLE J
PHLOEM K
XYLEM L
LATERAL ROOT GROWTH M

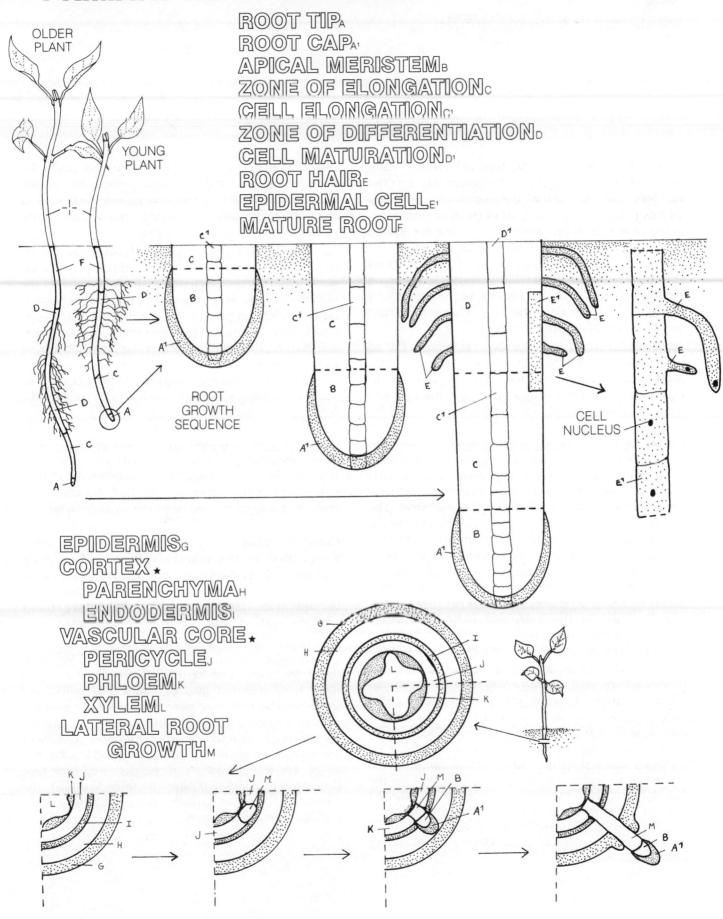

62
SHOOT GROWTH

Color the active terminal bud (A) and existing leaves (C) on the diagram of a young plant at the far left and the apical meristem (B), existing leaves (C), and maturing stem (J) in the enlargement of the terminal bud. Also color the apical meristem (B), the existing leaves (C), first leaf pair (D), second leaf pair (E), and primordia of the third leaf pair (F), axillary buds (I), maturing stem (J), and the diagrammatic sequence of cell elongation (K) in the first four diagrams, from left to right, along the bottom.

An active *terminal bud* at a shoot tip consists of an *apical meristem* in the region of the stem apex and a portion of *maturing stem* bearing a few pairs of *existing leaves.* This plate follows primary shoot growth from the *apical meristem* at the shoot tip through the production of five leaf pairs. The diagram at the far left on the bottom of the plate shows the position of the *apical meristem* at the beginning of the sequence. As growth proceeds, the leaf primordia of the *first leaf pair* (for the growth sequence illustrated) are produced in the region of the *apical meristem.* The *apical meristem* provides a basis for growth by the continuous addition of both new meristematic cells and cells that will enter the maturation process. A primary factor in stem growth is cell elongation in the direction of stem growth. The shoot elongates primarily by *cell elongation* of the cells in the several internode areas in the *maturing stem* immediately below the *apical meristem.*

Early in the development of the *first leaf pair,* one or more *axillary buds* develop at their base above their juncture with the *maturing stem.* As the *apical meristem* grows upward, due to a combination of new cell production and *cell elongation,* new leaf primordia as well as *axillary buds* develop. Because *cell elongation* in the internode areas in the *maturing stem* just below the shoot tip has just begun, the pairs of immature leaves remain closely clustered. In the fourth figure from the left in the sequence across the bottom of the plate, *cell elongation* in the internode area between the *first leaf pair* and *second leaf pair* is well underway, and leaf primordia for the *third leaf pair* are formed.

Color the previously listed features and the fourth leaf pair (G) and leaf primordia of the fifth leaf pair (H) on the remaining diagrams of the primary growth sequence across the bottom of the plate.

Continued primary shoot growth leads to the formation of additional stem material and leaf pairs. Internode length is established by *cell elongation* in the *maturing stem,* which provides for further separation of older leaf pairs. From the time of their beginning as leaf primordia, the leaves continue to develop and mature, eventually to attain the form and size of mature leaves.

In the final diagram of the primary shoot growth sequence, the *fourth leaf pair* and leaf primordia for the *fifth leaf pair* have formed. The *first leaf pair* has matured, and *cell elongation* in the internode area between the *first leaf pair* and *second leaf pair* is almost complete. Below the *first leaf pair* is *mature stem* in which *cell elongation* is completed and final internode length has been established. No additional growth in length will occur in the *mature stem.*

Color the diagram of the older plant following maturation of the maturing stem and existing leaf pairs from the young plant and formation of new primary shoot growth by the activities of the apical meristem. Also color the enlargement of the terminal bud.

Maturation of the immature shoot portion of the younger plant, at left, has occurred in the diagram of the older plant. The *existing leaves* on the young plant have matured into two new leaf pairs, and leaves of the *first leaf pair* formed in the primary growth sequence across the bottom of the plate are also mature. The active *terminal bud* now consists of the *apical meristem,* at a higher position on the shoot, the leaf primordia of the *fifth leaf pair* (not shown), the *fourth leaf pair,* the *third leaf pair,* the *second leaf pair,* the *axillary buds* (not shown), and a portion of *maturing stem.*

PRIMARY SHOOT GROWTH.

TERMINAL BUD A
APICAL MERISTEM B
EXISTING LEAVES C
LEAF PRIMORDIA ★
1st LEAF PAIR D
2nd LEAF PAIR E
3rd LEAF PAIR F
4th LEAF PAIR G
5th LEAF PAIR H
AXILLARY BUD I
MATURING STEM J
CELL ELONGATION K
MATURE STEM L

YOUNG PLANT

OLDER PLANT

63
ROOT GROWTH

Color the root cap (A), protoderm (B¹), ground meristem (C¹), and procambium (D¹) on the lowermost root section (root tip) depicting the root tip meristematic region.

Initial root formation is due to the production of new cells, which remain meristematic, by a root apical meristem (not shown) that is located at the root tip. A protective *root cap,* which covers the delicate apical meristem like a thimble, is also produced by the apical meristem. In the meristematic region immediately behind (above) the apical meristem, the initial formation of the primary meristems of the three tissue systems of the plant body (root) occurs. The meristematic regions of the plant body are the dermal tissue system, the fundamental tissue system, and the vascular (fascicular) tissue system. The meristematic tissue of the dermal tissue is the *protoderm,* the outermost meristematic region of the root. The ground (fundamental) tissues are formed by the *ground meristem;* the vascular (fascicular) tissues are formed by the *procambium.* Continued cell division within these three meristematic regions produces additional new cells as a basis for growth.

Color the epidermis (E), cortex parenchyma (F), and endodermis (G) of the cortex and the pericycle (H), primary phloem (I), and primary xylem (J) of the vascular region on the second root section depicting primary root growth.

The *protoderm* produces the *epidermis,* a protective covering for the entire primary plant body that is usually no more than one cell layer thick. During secondary growth, the *epidermis* is disrupted. The *ground meristem* of the root produces the cortex, which consists of an outer, often thick, area of *cortex parenchyma* and an inner layer, called the *endodermis,* that is usually a single cell layer thick. Cells of the *endodermis* secrete a substance into their walls that is impervious to water movement. This substance is secreted only into the portions of the walls that are in contact with other *endodermis* cells. These water-impervious, gasketlike thickenings, called casparian strips (not shown), form a barrier to water movement. Thus, water movement through the *endodermis* must

be either directly through the outer walls, protoplast, and inner walls of the *endodermis* cells or through specialized areas, opposite the arms (lobes) of the *primary xylem,* where no water-impervious substance is produced. In this way, the *endodermis* controls water movement into the plant. *Endodermis* is typically formed only in roots. The *procambium* of the root produces the *pericycle, primary phloem,* and *primary xylem.* The *primary xylem* is typically star-shaped, and the bundles (strands) of *primary phloem* are located between the primary xylem arms.

Color the previously listed features and the vascular cambium (K), secondary phloem (I¹), and secondary xylem (J¹) on the uppermost root section depicting secondary growth. Do not color the components of the pie-shaped wedge of root.

Woody plants enlarge due to secondary growth. Tissue areas of a root undergoing secondary growth are, from the outside to the center, the *epidermis,* the cortex, and the vascular core. Secondary growth begins with the formation of a *vascular cambium* between the *primary phloem* and *primary xylem.* *Secondary phloem* is produced on the outside of the *vascular cambium; secondary xylem* is produced on the inside. *Secondary phloem* and *secondary xylem* are formed as long as the *vascular cambium* remains active (for the life of the plant).

Color the dermal tissue (B), fundamental tissue (C), and vascular tissue (D) regions in the pie-shaped wedge previously left uncolored.

In review, the apical meristem is the original source of the primary plant body tissues, and three tissue systems occur in a vascular plant body. In the root, the *dermal tissue* consists of only the *epidermis;* the *fundamental tissues* consist of the *cortex parenchyma* and *endodermis;* and the *vascular tissues* consist of the *pericycle, primary phloem,* and *primary xylem.* In plants undergoing secondary growth, the *vascular tissues* also include the *vascular cambium, secondary phloem,* and *secondary xylem.*

ROOT ANATOMY AND SECONDARY GROWTH.

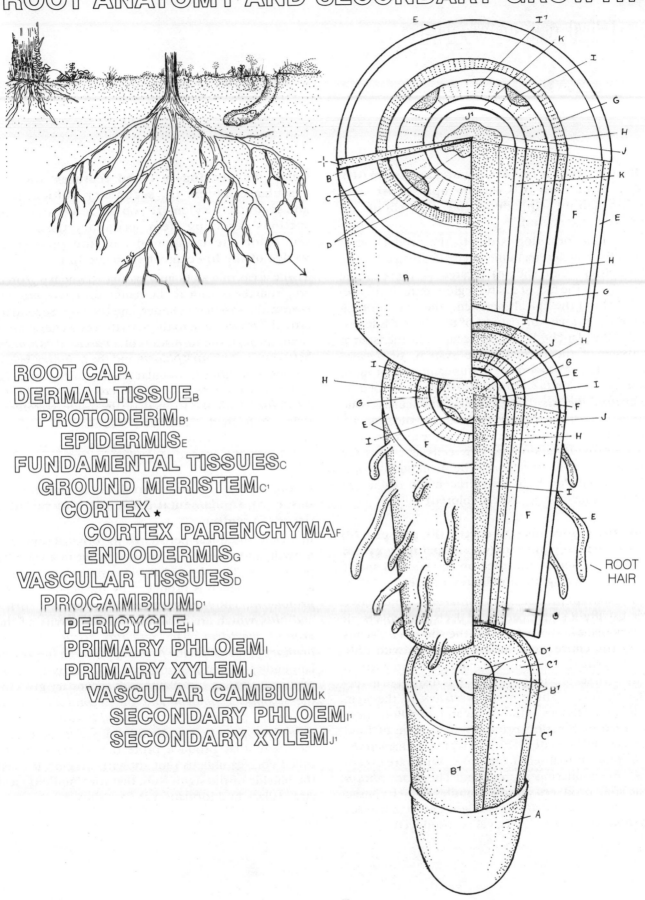

ROOT CAP A
DERMAL TISSUE B
    PROTODERM B¹
    EPIDERMIS E
FUNDAMENTAL TISSUES C
    GROUND MERISTEM C¹
        CORTEX ★
            CORTEX PARENCHYMA F
        ENDODERMIS G
VASCULAR TISSUES D
    PROCAMBIUM D¹
        PERICYCLE H
        PRIMARY PHLOEM I
        PRIMARY XYLEM J
        VASCULAR CAMBIUM K
        SECONDARY PHLOEM I¹
        SECONDARY XYLEM J¹

ROOT
HAIR

64
STEM GROWTH

Color the protoderm (A¹), ground meristem (B¹), and procambium (C¹) on the uppermost stem section (shoot tip) depicting the shoot tip meristematic region.

Initial shoot formation is due to the production of new cells, which remain meristematic, by a stem apical meristem (not shown) located at the extreme shoot tip. In the meristematic region immediately behind (below) the apical meristem, the initial formation of the primary meristems of the three tissue systems of the plant body (shoot) occurs. The meristematic tissue of the dermal tissue system, or *dermal tissue,* is the *protoderm.* The fundamental (ground) tissue system, or *fundamental tissues,* is formed by the *ground meristem;* the vascular (fascicular) tissue system, or *vascular tissues,* is formed by the *procambium.* Continued division in these three meristematic regions provides additional new cells as a basis for growth in these stem regions. The small budlike projections from the shoot tip meristematic region are leaf primordia (not separately colored).

Color the epidermis (D), cortex (E), and pith (F) and the primary phloem (G) and primary xylem (H) of the vascular bundles on the second shoot section depicting stem primary tissue.

The *protoderm* produces the *epidermis,* a protective layer usually a single cell layer deep that covers all primary plant body surfaces. In the shoot, *epidermis* covers the entire stem surface in plants with only primary growth but only the younger stem parts in plants with secondary growth, as it is disrupted in secondary growth. The *ground meristem* of the stem produces the tissues of the *cortex* and *pith.* Both these stem regions function primarily in storage of food materials and/or water. The pith region is characteristic of stems; it is absent from roots. With rare exceptions, no endodermis is produced. The *procambium* of the stem produces vascular bundles, which consist of an outer area of *primary phloem* and an inner area of *primary xylem.* No pericycle is produced.

Color the previously listed features and the vascular cambium (I), secondary phloem (G¹), and secondary xylem (H¹) on the lower two stem sections depicting stem secondary tissues. Do not color the components of the pie-shaped wedge of the lowermost stem section.

Tissue areas of a stem undergoing secondary growth are, from the outside to the center, the *epidermis,* the *cortex,* the vascular cylinder, and the *pith.* Secondary growth, growth due to the activities of a lateral meristem, through the formation of a *vascular cambium* between the *primary phloem* and the *primary xylem,* produces additional vascular tissues in the vascular cylinder. On the outside of the *vascular cambium, secondary phloem* is formed; on the inside, *secondary xylem* is formed. Continued secondary growth, as shown in the last stem section, produces additional *secondary phloem* and *secondary xylem.*

In the wedge left uncolored, color the dermal tissue (A), fundamental tissue (B), and vascular tissue (C) regions.

In review, the apical meristem is the original source of growth, and three tissue systems occur in a vascular plant body. The *dermal tissue* consists only of the *epidermis,* which is produced by the activity of the *protoderm.* The *fundamental tissues* are the *cortex* and *pith,* which are produced by the activity of the *ground meristem.* The *vascular tissues* are the *primary phloem* and the *primary xylem* in the vascular bundles, which are produced by the activity of the *procambium,* and in plants with secondary growth, the *secondary phloem* and the *secondary xylem,* which are produced by the *vascular cambium.* As secondary growth commences, a series of lateral meristems that produce protective layers of cork, called cork cambiums (not shown), develop toward the outside of the stem. Note that for simplicity, leaf and axillary bud formation are not shown.

STEM ANATOMY AND GROWTH.

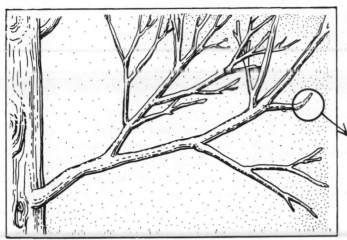

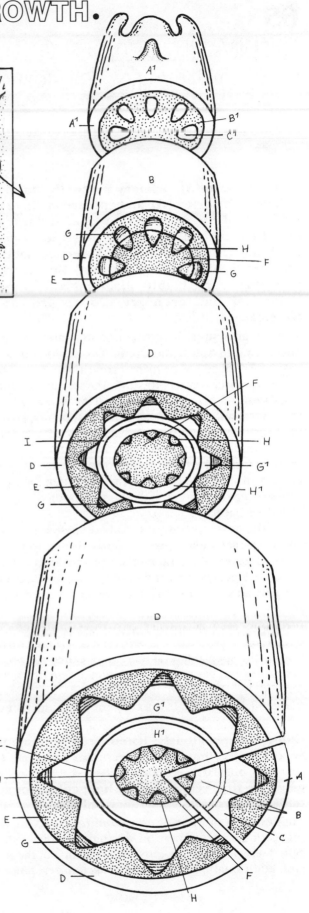

DERMAL TISSUE A
 PROTODERM A¹
  EPIDERMIS D
FUNDAMENTAL TISSUES B
 GROUND MERISTEM B¹
  CORTEX E
  PITH F
VASCULAR TISSUES C
 PROCAMBIUM C¹
  PRIMARY PHLOEM G
  PRIMARY XYLEM H
 VASCULAR CAMBIUM I
  SECONDARY PHLOEM G¹
  SECONDARY XYLEM H¹

65
ORGANIZATION OF SECONDARY GROWTH

Color the pith (A), primary xylem (B), primary phloem (C), cortex (D), and epidermis (E) on the one-fifth of the stem section labeled "primary growth" and these features plus the vascular cambium (F) on the one-fifth section labeled "beginning of secondary growth." Bear in mind that these are highly diagrammatic presentations of the secondary growth process in the stem.

Primary growth is the growth of stem and root produced by the apical meristem. The regions of stem tissue produced by primary growth are, from stem center to outer surface, the *pith,* the vascular cylinder consisting of *primary xylem* and *primary phloem,* the *cortex,* and the *epidermis.* In most plants, growth in length is due only to the activities of the apical meristems at stem and root tips. Thus, primary growth forms the basic plant body and accounts for growth in stem and root length. Plants that have only primary growth are relatively soft bodied and are called herbaceous.

For the development of woody tissues, as in shrubs and trees, secondary growth (growth due to the activity of a special lateral meristem) must occur. Thus, secondary growth forms woody tissues and accounts for growth in stem and root diameter. The process of secondary growth is similar in stems and roots; therefore, only stem secondary growth is illustrated. The beginning of secondary growth starts with the formation of the special lateral meristem, called the *vascular cambium,* between the *primary xylem* (the xylem formed by the apical meristem) and the *primary phloem* (the phloem formed by the apical meristem).

Color the previously listed features and the secondary xylem (G) and secondary phloem (H) on the one-fifth section labeled "secondary growth—first year." Choose light colors for secondary xylem and secondary phloem.

The function of the *vascular cambium* is to provide for secondary growth by producing additional xylem, called *secondary xylem,* toward the inside of the stem and additional phloem, called *secondary phloem,*

toward the outside of the stem. By the end of the first year of secondary growth, there is a layer of *secondary xylem* formed between the *primary xylem* and the *vascular cambium.* Similarly, there is a layer of *secondary phloem* formed between the *vascular cambium* and the *primary phloem,* which is outermost.

Color the previously listed features on the two one-fifth sections labeled "second year" and "third year." Note that the numbers within the rings (arcs) of secondary xylem and secondary phloem correspond to the year in which they were produced. Also color the parenchyma rays (I) in these sections.

During each growth season, additional increments of *secondary xylem* and *secondary phloem* are added to the stems, resulting in an ever-increasing diameter. Because the *vascular cambium* produces each season's increment of *secondary xylem* and *secondary phloem,* the youngest layers of these two secondary tissues are those immediately adjacent to the *vascular cambium.* Thus, the oldest *secondary xylem* is that nearest the center of the stem, and the oldest *secondary phloem* is nearest the outside. As the *vascular cambium* forms additional *secondary xylem,* it moves outward in the enlarging stem.

The older layers of *secondary phloem,* produced when the stem had a smaller diameter, are pushed outward by the formation of additional increments of *secondary xylem* and *secondary phloem.* The older *secondary phloem* cannot form additional *secondary phloem* (only the vascular cambium forms *secondary phloem*) to accommodate the ever-increasing stem diameter. As a result, the older cylinders of *secondary phloem* stretch and split to form fissures that are filled by a proliferation of parenchyma cells to form broad *parenchyma rays.*

To a degree, the *epidermis* and *cortex* can accommodate the increase in stem diameter by cell division, but they, too, eventually split to form fissures in the bark.

THE VASCULAR CAMBIUM.

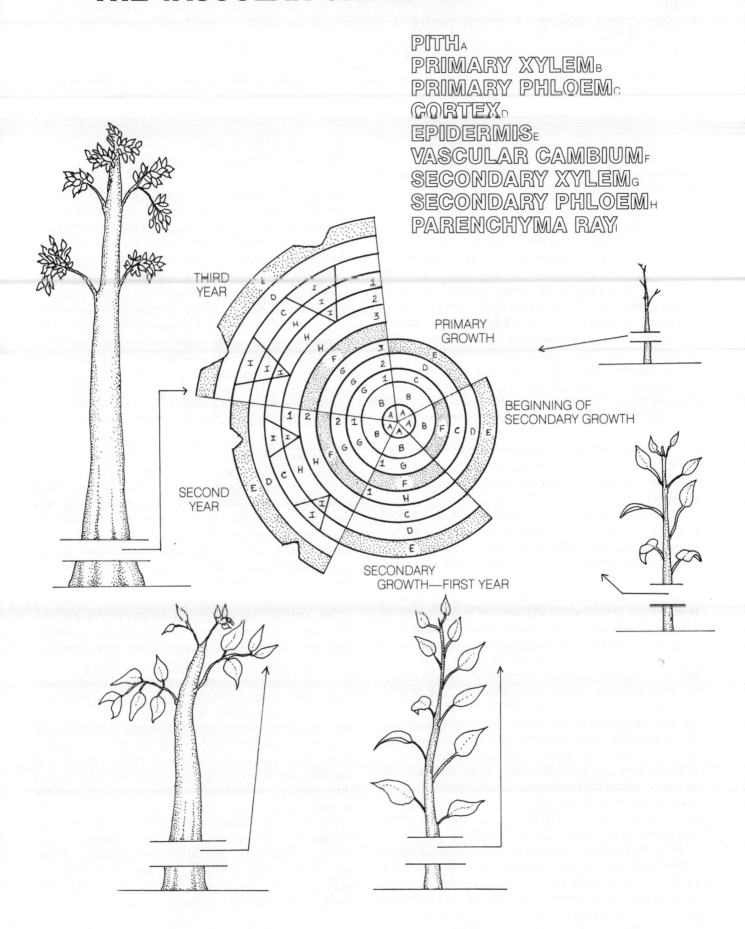

PITH A
PRIMARY XYLEM B
PRIMARY PHLOEM C
CORTEX D
EPIDERMIS E
VASCULAR CAMBIUM F
SECONDARY XYLEM G
SECONDARY PHLOEM H
PARENCHYMA RAY I

THIRD YEAR

PRIMARY GROWTH

SECOND YEAR

BEGINNING OF SECONDARY GROWTH

SECONDARY GROWTH—FIRST YEAR

Color the vascular cambium (A) and phloem (B), phloem fibers (C), cork cambium (D), cork (E), and phelloderm (F) of the bark on the pie-shaped stem section at the right.

The bark of a tree consists of all the tissues external to the *vascular cambium.* Most of the bark consists of *phloem,* especially secondary phloem. Distributed within the *phloem* are bundles of *phloem fibers* that add strength. The youngest *phloem,* that toward the *vascular cambium,* is the active *phloem* that functions in food material transport. As a stem enlarges in diameter, through the activity of a *vascular cambium* in secondary growth, the outer layers, including the epidermis (not shown) of the primary plant body, become stretched as the stem circumference increases. Eventually, these outer tissues split to form cracks or fissures. This diminishes their efficiency as a protective layer. To counteract this loss of protection, plants form a second type of lateral meristem called phellogen or *cork cambium* to form protective layers internal to the stem surface. While a *cork cambium* remains active, it can accommodate the increasing stem diameter.

The tissues formed by the *cork cambium* are *cork* or phellem toward the outside of the stem and *phelloderm* toward the inside. The *cork cambium, cork,* and *phelloderm* are collectively called the periderm. During development, the *cork* cells secrete substances impervious to water movement into their walls. This is why *cork* is useful as a closure for wine bottles. At maturity, the cork cells are dead. Thus, *cork* consists only of cell walls when fully mature. Since the *cork* is impervious to water, substances essential to life, such as food materials and water, cannot move from internal stem areas to the tissues outside the *cork.* Within the active *phloem,* the splits that form due to growth are filled by parenchyma cells. Since the tissues outside the most internal periderm are all dead, no filling occurs. In trees that do not soon shed older layers of dead bark (such as most pines) deep fissures may develop, often in a pattern characteristic of the species. In smooth-barked trees, the tissues outside the *cork* are shed, in strips (eucalyptus) or plates (sycamore or plane tree), as dead bark.

The first *cork cambium* to develop typically forms within the cortex (not shown). Several successive periderms are usually formed during the life of a tree, and the dead bark, the bark outside the most internal periderm, may consist of one to several nonfunctional (dead) periderms as well as inactive (dead) *phloem.* Only one periderm is illustrated.

Color the 1st year (G), 2nd year (H), 3rd year (I), and 4th year (J) of growth of the wood tissue (xylem) on the tree and the 5th year (K), 6th year (L), and 7th year (M) of growth of wood tissue on the tree and on the pie-shaped stem section. Also color the sections of stem and root.

Because the *vascular cambium* adds increments of new wood (the secondary xylem) to the layers of older wood, the oldest wood is toward the center of the stem and the youngest wood, toward the outside of the stem. In the seven-year-old stem illustrated, the *1st year xylem,* produced during the first season of growth, is present as a small cone, equivalent to the height of the tree at that time, at the center of the base of the stem. During the second year of growth, primary growth further elongates the stem while secondary growth produces a new layer of secondary xylem, *2nd year xylem,* over the cone of *1st year xylem.* The secondary phloem is not shown. While primary growth elongates the main stem and branches, increments of secondary xylem and secondary phloem produced each succeeding year add to the stem diameter. During each season, more secondary xylem than secondary phloem is produced by the *vascular cambium.* This, coupled with the retention of all secondary xylem formed and the loss of older layers of secondary phloem and primary tissue, produces a stem with abundant wood and relatively little bark. Continued secondary growth in successive seasons produces the *3rd year xylem, 4th year xylem, 5th year xylem, 6th year xylem,* and *7th year xylem.* Each year's growth of xylem can be recognized by the presence of a growth ring (indicated by the dark lines separating each year of xylem growth), which are due to a change in the type and size of xylem cells produced by the *vascular cambium* as growth slows at the end of the growing season and the plant enters dormancy. Growth rings are areas of dense xylem due to the presence of numerous small cells.

On the diagrammatic tree, two branches are illustrated. The lowermost branch began growth in the third year but stopped growth in the fourth year. The base of this branch is becoming covered by additional annual increments of secondary xylem (wood). The uppermost branch is actively growing.

THE CORK CAMBIUM AND ANNUAL GROWTH.

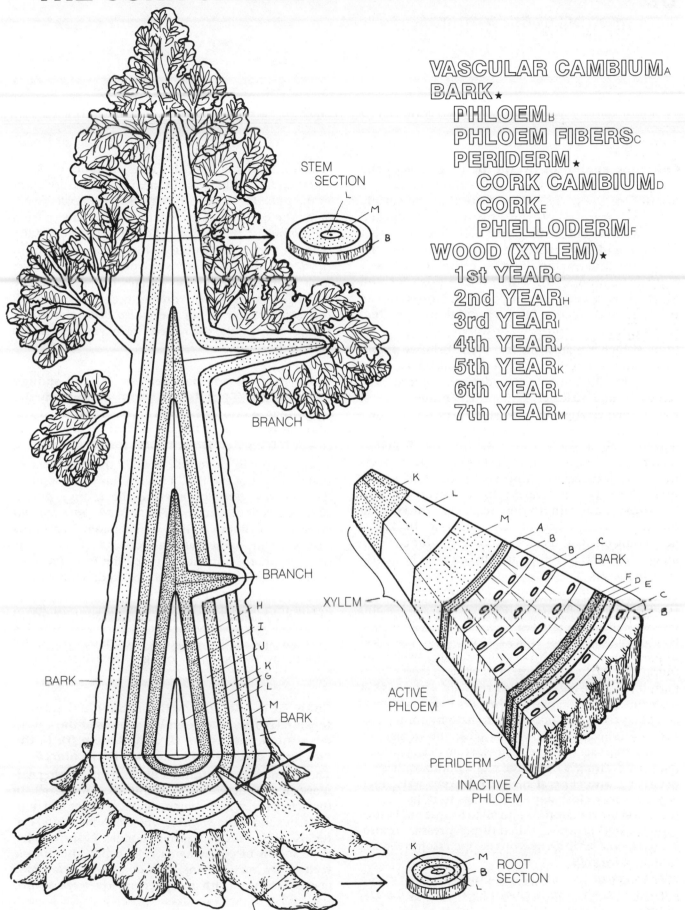

VASCULAR CAMBIUM A
BARK ★
　PHLOEM B
　PHLOEM FIBERS C
PERIDERM ★
　CORK CAMBIUM D
　CORK E
　PHELLODERM F
WOOD (XYLEM) ★
1st YEAR G
2nd YEAR H
3rd YEAR I
4th YEAR J
5th YEAR K
6th YEAR L
7th YEAR M

STEM SECTION

BRANCH

BRANCH

BARK

BARK

BARK

ROOT SECTION

XYLEM

ACTIVE PHLOEM

PERIDERM

INACTIVE PHLOEM

67
LEAF STRUCTURE

Color the epidermis (A) on the diagrams of the whole leaf, leaf section, and enlargement of a portion of the leaf section. Note that the cuticle (B) is separately colored only on the enlarged portion. Also color the guard cells (C) on the leaf section and the enlarged portion and on the diagrams illustrating their function at the lower right. The stoma (D), which is the pore formed by the guard cells, is not colored. Note that the fundamental system and vascular system features remain uncolored for now.

In most higher plants, the leaf is the primary photosynthentic structure (organ). A generalized dicot leaf is illustrated. Though the dermal, fundamental (ground), and vascular systems are found in most leaves, their arrangement and structure vary. The dermal system consists of a single cell layer, the *epidermis,* and a waxy layer, the *cuticle,* which is secreted by the *epidermis.* The dermal system covers the entire leaf surface to provide protection against desiccation. The epidermal cells are usually cuboidal in cross section, with the outer cell walls thicker than the inner walls. The *cuticle* varies in thickness, depending on plant species and aridity of the environment. *Cuticle* thickness is usually greatest in an arid environment.

Since the *epidermis* and *cuticle* effectively inhibit the movement of water and gases through the leaf surface and since gas exchange is necessary for photosynthesis, a means of allowing gases to pass into and out of a leaf is required. Numerous stomates, which are found only in the lower leaf *epidermis* in most plants, permit gas exchange and also regulate water loss. A stomate consists of two *guard cells* that function as gatelike valves that open the stomate by forming an opening, called a *stoma,* or that close the stomate by coming together. In most plants, stomates open during the day, for gas exchange while photosynthesis is occurring, and close at night to prevent water loss. Stomates may close during hot days to reduce water loss. Stomate function is controlled by internal hydrostatic (water) pressure, called turgor pressure (represented by the small arrows on the open stomate),within in the *guard cells.*

Each *guard cell,* unlike the other cells of the *epidermis,* contains several chloroplasts. During the day, photosynthesis within the chloroplasts in the *guard cells* produces sugars. This causes an influx of water and therefore an increase in the turgor pressure within the *guard cells.* As a result, the *guard cells* swell. Since the outer, thin-walled areas (those away from the *stoma*) of the *guard cells* distend more readily than the inner, thick-walled areas (those adjacent to the *stoma*), the *guard cells* swell outward and pull the thick walls outward. This creates the opening, the *stoma,* of the stomate. At night, the sugars are used; the excess water leaves the *guard cells;* the turgor pressure drops; the *guard cells* shrink; and the *stoma* is closed.

Color the palisade parenchyma (E) and spongy parenchyma (F) cells of the mesophyll on the large leaf section diagram in the middle of the plate and on the enlarged section of leaf in the lower left corner. Do not color the air spaces (G).

The mesophyll, or middle leaf, is the fundamental tissue system of the leaf. Two regions of mesophyll, the *palisade parenchyma* and the *spongy parenchyma,* can be recognized. The *palisade parenchyma* consists of cylindrical, thin-walled parenchyma cells that contain numerous chloroplasts. This is the primary photosynthetic tissue of the leaf. *Palisade parenchyma* cells are closely packed, with only a very small air space between cells.

The thin-walled cells of the *spongy parenchyma* are irregularly shaped and loosely packed, forming large air spaces that create a continuous *air space* within the mesophyll. Most gas exchange, O_2 and CO_2, occurs in this area.

Color the bundle sheath (H) and bundle sheath extension (H¹) and the complex tissues of the vascular system, the xylem (I) and phloem (J), in the two large diagrams that are partially colored.

The vascular system forms a diversely branching network of conductive veins within the mesophyll. In the veins, the *xylem,* which conducts water and minerals, lies above the *phloem,* which conducts food materials. Each vein, or vascular bundle, is surrounded by a sheath of parenchyma cells called the *bundle sheath.* On major veins, *bundle sheath extensions* connect the vein with the upper and lower *epidermis.*

GENERAL LEAF ANATOMY.

DERMAL SYSTEM★
  EPIDERMIS_A
  CUTICLE_B
STOMATE★
   GUARD CELL_C
   STOMA_D-¦-

FUNDAMENTAL SYSTEM★
  MESOPHYLL★
  PALISADE PARENCHYMA_E
  SPONGY PARENCHYMA_F
  AIR SPACE_G-¦-

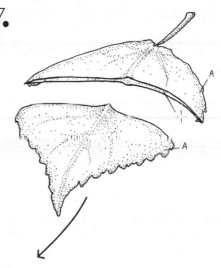

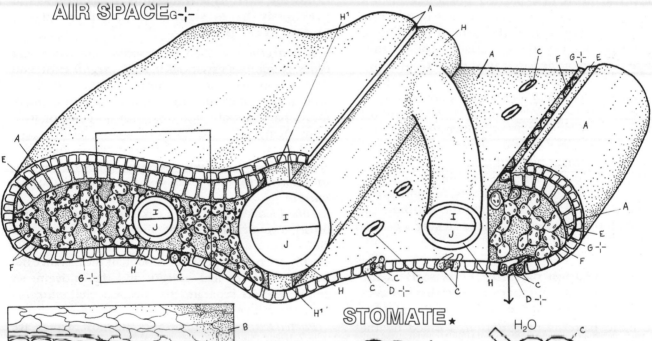

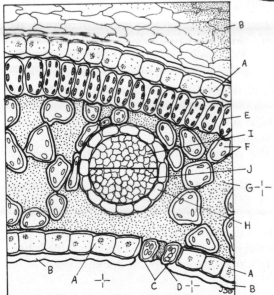

STOMATE★

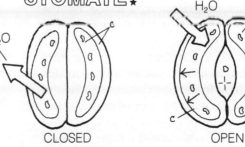

CLOSED   OPEN

BUNDLE SHEATH_H
BUNDLE SHEATH EXTENSION_{H¹}
VASCULAR SYSTEM★
  XYLEM_I
  PHLOEM_J

Color the leaf (A), its major morphological regions, the lamina (B) and the petiole (C), and the stem (D) and axillary bud (E) on the three diagrams at the top of the plate.

Leaves come in a variety of shapes and sizes. Most simple *leaves* can be divided into two morphological regions. The broad, or expanded, portion of a *leaf* is called the *lamina,* or blade, in reference to its broad, thin shape. In most *leaves,* the *lamina* functions as the primary photosynthetic surface. The narrow, stalklike portion of the *leaf,* called the *petiole,* which is located between the *lamina* and *stem,* often functions to position the *leaf* for efficient photosynthesis. The portion of the *petiole* near the stem is called the leaf base (not separately colored). Since all *leaves* have one or more *axillary buds* on the *stem* immediately above the attachment point of the *petiole,* in the axil of the *leaf,* this feature serves as a diagnostic trait for a single *leaf.* The *leaves* of some plants, or some *leaves* on some plants, lack a *petiole,* and the *lamina* is attached directly to the *stem. Leaves* lacking a *petiole* are called sessile *leaves.*

Color the venation patterns (F) as well as the previously listed structures on the three leaves labeled ''venation'' in the middle of the plate.

Three basic patterns of *venation,* formed by the distinctive arrangement of major veins, are commonly encountered. *Leaves* with *pinnate venation* have a single, centrally positioned, main vein running from the base of the *lamina* to its tip. Numerous secondary veins branch at an angle from the central main vein to produce a featherlike pinnate pattern of major veins.

If a few to several major veins arise from a common point near the base of the *lamina* and radiate outward toward the *leaf margin,* often to the tips of lobes, like fingers from a palm, the venation pattern is called *palmate* or digitate.

A third commonly encountered pattern is *parallel venation,* in which several major veins arise from the base of the *lamina,* diverge slightly, and then run roughly parallel for most of the length of the *lamina*

before converging toward the leaf tip. This pattern is characteristic of a group of flowering plants called monocots. Numerous small *parallel veins* are typically present (only a few are diagrammatically illustrated).

Color the various leaf margins (G) as well as the previously listed structures illustrated at the bottom of the plate.

Leaf *margins* are used to describe and name leaf types. Numerous variations in leaf *margins* exist, and though only six types are illustrated, they provide a good sample of commonly encountered types. *Entire leaves* have smooth *margins* with no lobes or incisions. Many different forms of *serrate leaves* exist, but the basic pattern is a leaf *margin* with roughly triangular, sharp toothlike projections, or serrations, of the *lamina* that are evenly spaced around the leaf *margin.*

Lobed leaves, as in some oaks, have *margins* with broad, rounded lobes that alternate with rounded clefts that are irregularly spaced. The incised clefts usually are not deeper than about half the distance between the outer leaf *margin* and the midvein; so ample *lamina* separates the deepest cleft *margins* from the midvein.

Parted leaves, as in some oaks and most maples, are divided into several sections by angular, often deep, clefts in the *margins.* The clefts may approach the major veins but do not touch them; so lamina material separates the deepest cleft *margins* from the major veins.

Palmatifid leaves, as in the rice paper plant, have *palmate venation* and deeply incised *margins.* The clefts of the *margins* alternate with the fingerlike lamina lobes that surround the major veins. Though deeply incised, the clefts do not reach the common origin point of the main veins.

Pinnatifid leaves, like *palmatifid leaves,* have deep clefts in the *margin,* but the clefts follow a pinnate venation pattern so that a series of lamina lobes, arranged in a *pinnate pattern,* is present.

BASIC LEAF STRUCTURE.

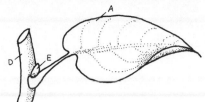

LEAF_A
　LAMINA_B
　　PETIOLE_C
STEM_D
AXILLARY BUD_E

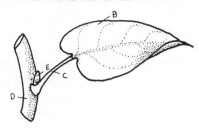 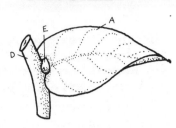

PETIOLATE ★　　　　**SESSILE** ★

VENATION_F

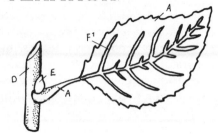 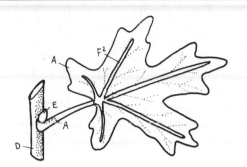 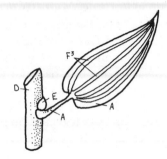

PINNATE_{F1}　　　**PALMATE**_{F2}　　　**PARALLEL**_{F3}

MARGINS_G

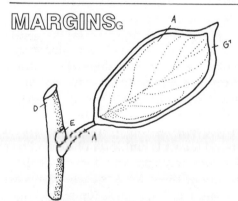 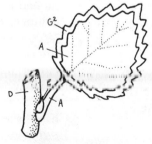 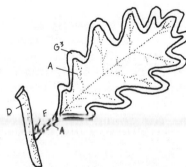

ENTIRE_{G1}　　**SERRATE**_{G2}　　**LOBED**_{G3}

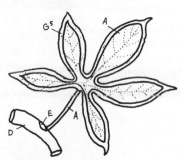

PARTED_{G4}　　**PALMATIFID**_{G5}　　**PINNATIFID**_{G6}

Color the alternate (A), opposite (B), and whorled (C) leaf arrangements and the stems (D) and axillary buds (E) on the diagrams of leaf arrangements at the top of the plate.

Three leaf arrangements, based on the number of leaves present at each node, are recognized. If a single leaf is present at each node, the leaf arrangement is *alternate*. Roses and many other plants have alternate leaves. The leaves, each with one or more axillary buds, are attached to the *stem* in an ascending pattern of spirals about the *stem*. Plants with two leaves present at each node have an *opposite arrangement* of leaves. Two leaves occur opposite one another at each node. Many plants, including maples and coleus, have *opposite* leaves. The presence of three or more leaves at each node, as in Easter lilies, is called a *whorled arrangement* because the leaves appear in whorls at each node. Because of the presence of *axillary buds*, branching patterns usually follow patterns of leaf arrangements.

Color the compound leaf types and the components of a compound leaf, the leaflets (F) and rachis (G), if present, and petiole (H). Note that each type, except bipinnate (K) and ternate (L), is illustrated twice. Color one illustration to show the components and the other with the color of the compound leaf type: odd pinnate (I), even pinnate (J), or palmate (M). Note that in the case of bipinnate (K) and ternate (L), the title only receives a separate leaf-type color.

The blade portion of compound leaves is divided into two or more subdivisions, called *leaflets,* that closely resemble individual leaves. However, a true leaf can always be determined by the presence of one or more *axillary buds* at its point of attachment to the *stem*. Individual *leaflets* never have an *axillary bud* at their base, though the compound leaf they form does have one or more *axillary buds*. In many compound leaves, the midvein forms a narrow, stemlike structure, called the *rachis,* to which the *leaflets* are attached. Thus, without careful observation, *leaflets* appear to be true leaves attached to a true *stem*. The diagnostic feature of a true leaf is the presence of one or more *axillary buds* above its point of attachment. Most compound leaves have a distinct *petiole*. *Leaflets* may be sessile on the *rachis* or attached to it by a petiolelike stalk.

Many arrangements of *leaflets* in various compound leaves occur, and only a few common types are described here. Pinnately compound leaves have opposite leaflet pairs arranged in a pinnate (featherlike) manner. *Odd pinnate* compound leaves, as in roses, have the *rachis,* or midvein, terminated by a single *leaflet* and therefore an odd number of *leaflets* present on the leaf. In some plants, such as vetch *(Vicia),* the rachis tip is modified into a grasping tendril. In *even pinnate* compound leaves, such as in silk-tassle tree *(Albizia),* the leaf is terminated by a pair of *leaflets* so that the leaf has an even number of *leaflets* present.

In *bipinnate* compound leaves, the pinnately arranged leaf divisions are further divided in a pinnate manner. The result is a pinnate arrangement of *leaflets* on pinnately arranged branches of the *rachis*. *Ternate* compound, or trifoliate, leaves, such as found on clover, have three *leaflets*. The overall pattern for ternately compound leaves is multiples of three. For example, a twice ternately compound leaf (not shown) has three clusters of three *leaflets* each. Plants with *palmate* compound leaves, such as lupines, have four or more *leaflets,* all attached to the tip of a true *petiole* at one common point of origin. This arrangement resembles the palm of the hand with the fingers analogous to *leaflets,* hence the name *palmate*. Another name used for this compound leaf type is digitate, in reference to the similarity to the fingers of the hand.

LEAF ARRANGEMENTS AND COMPOUND LEAVES.

ARRANGEMENT⋆

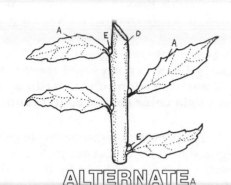

ALTERNATEA

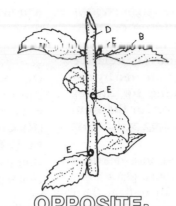

OPPOSITEB

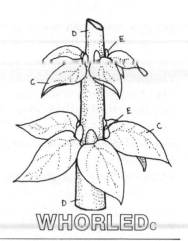

WHORLEDC

COMPOUND LEAVES ⋆

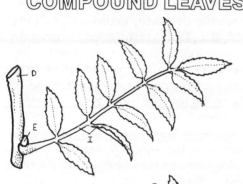

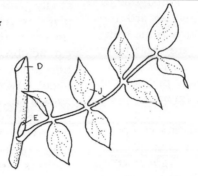

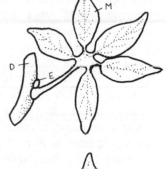

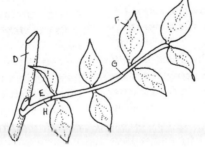

ODD PINNATEI

EVEN PINNATEJ

PALMATEM

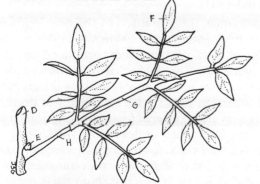

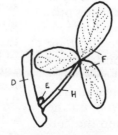

BIPINNATEK

TERNATEL

STEMD
AXILLARY BUDE
LEAFLETF
RACHISG
PETIOLEH

On the diagrams labeled "phototropism," color the arrows indicating the direction of the light source (A), stem (B), shoot tip (C), leaves (D), cell elongation (E), and the axillary buds (I). Relative plant hormone concentration is indicated by the degree of shading: darker shading for higher concentration. Choose a light color for the stem (B) and shoot tip (C) so the shading will show through.

Phototropism is a plant growth response due to a stimulus provided by a *light source*. Most plant *stems* exhibit a positive phototropic response to light. That is, the *stems* grow toward the *light source*. This effect can be observed in most houseplants, which, due to positive phototropism, tend to grow toward the nearest window. A few plant *stems*, such as the fruit-bearing stems of Kenilworth ivy, are negatively phototropic and grow away from the *light source*. Positive phototropism is an adaptation to position the photosynthetic surfaces in the most favorable position for efficient light reception.

If the primary *light source* is directly above a positively phototropic *stem* (as in the left diagram), the *stem* will grow vertically toward the *light source*. This response is due to plant growth hormones, especially one called auxin, produced by the *shoot tip* and immature *leaves*. (In the diagrams, the *shoot tip* is depicted as a simple dome for clarity.) These hormones regulate *cell elongation* in the zone of *cell elongation* and other aspects of plant growth not included here. When the *light source* is directly above the *shoot tip* (in line with the axis of the growing shoot), the hormones are evenly concentrated (shown by even shading) throughout the *shoot tip* and within any given level of the *stem's* zone of elongation below the *shoot tip*. Therefore, *cell elongation* at any given level is equal, and vertical growth of the *stem* directly toward the *light source* occurs.

If the *light source* is repositioned to one side of the *stem* (*light source* is shown repositioned to the right of the *stem* in the middle diagram), the hormones produced by the *shoot tip* become concentrated (shown by heavy shading) on the side opposite the *light source*. Since the hormones are transported only directly downward in the *stem*, hormonal concentration in the zone of *cell elongation* is also unequal. The result is greater growth, by *cell elongation*, on the side of higher hormone concentration and the curvature of the *stem* toward the light.

When the direction of *stem* growth is once again in line with the *light source* (as in the diagrams on the right), hormonal concentration once again becomes equal throughout the *stem*. *Cell elongation* (as in the tip of the far right diagram) is also once again equal

within the zone of *cell elongation*, and the *stem* continues growth directly toward the repositioned light source.

On the diagrams labeled "geotropism," color the previously listed features and the root (F), root tip (G), and direction of gravitational pull (H). Choose a light color for the root (F) and root tip (G).

Geotropism is a plant growth response due to gravitational pull (*gravity*). In plants that normally grow vertically, growth hormones produced by the *shoot tip* and, in the *root*, by the *root tip* are evenly distributed at any given level within the *stem* and *root*. If such a plant is repositioned in a horizontal position (as shown in the middle diagram), geotropic responses can be observed. If a plant is positioned horizontally, hormone concentration within *root* and *stem* become localized on the lower side of the *root* and *stem*. In the *root*, a positive geotropic response is observed as the *root* bends downward into the soil. This is apparently due to the inhibition of *cell elongation* in *root* cells by high concentrations of hormones. Also, growth inhibitors, produced by the *root tip*, transported upward, and then localized on the lower side of the *root*, may be involved.

In the *stem*, a negative geotropic response (growth away from the direction of gravitational pull) is observed as the *stem* bends upward away from the surface. As in a phototropic response, *stem* curvature is due to high concentrations of growth hormones, in this case, on the lower side of the *stem*, that enhance *cell elongation* in that area.

On the diagrams labeled "apical dominance," color the previously listed features and the lateral branches (J).

Apical dominance, the inhibition of the development of *axillary buds* into *lateral branches*, is due to the suppressive effects of hormones produced by the growing *shoot tip*. The degree of inhibition among plant species varies and accounts for some differences in growth form. In plants with strong apical dominance, there may be no *lateral branch* formation from *axillary buds* for several nodes below the *shoot tip*. The result is a well-developed, single-stemmed plant.

One means of overcoming apical dominance is by removing the *shoot tip*. Once the *shoot tip* is removed, its hormone production is eliminated. Hormone concentration in the *stem* diminishes, and a number of *axillary buds*, especially those nearer the cut *stem* tip, become active and begin growth.

PLANT HORMONE ACTION.

LIGHT SOURCE _A   ROOT _F
STEM _B   ROOT TIP _G
SHOOT TIP _C   GRAVITY _H
LEAF _D   AXILLARY BUD _I
CELL ELONGATION _E   LATERAL BRANCH _J

PHOTOTROPISM ★

GEOTROPISM ★

APICAL DOMINANCE ★

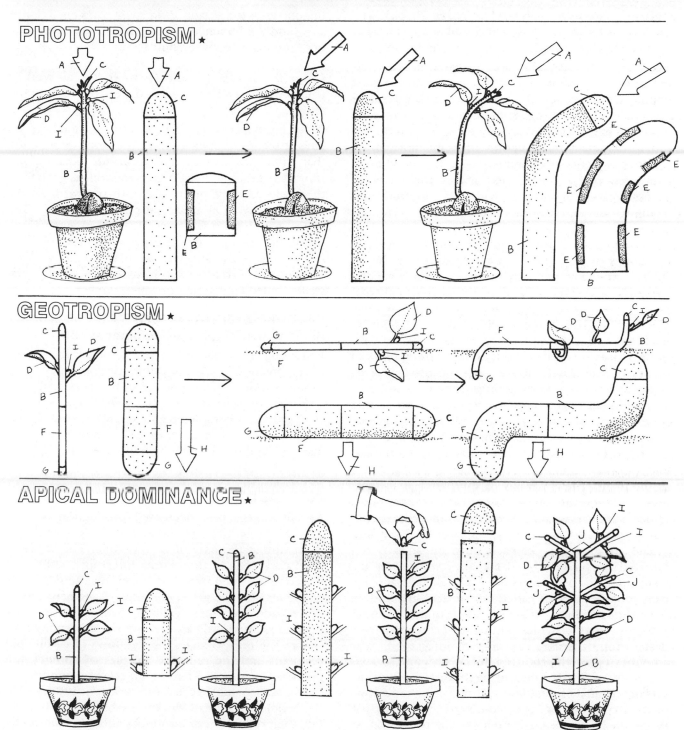

Color the stem (A), foliage leaf scars (B), axillary bud (C), terminal bud (D), and bud scale scars (E) on the shoot labeled "winter" at the far left.

The effects of seasonality are expressed in many different ways by plants. A few that are familiar to most people are dormancy, the period of active vegetative growth, and flowering. Some changes that may be observed in a deciduous woody dicot, such as the apple, provide an introduction to plant seasonality. Deciduous plants shed their leaves each year, whereas evergreen plants retain their leaves for a few to several years. This plate provides an opportunity to examine the external morphology of a deciduous *stem*.

During winter dormancy, the production of growth hormones by stem tips is minimized due to the influence of the winter environment. This is accompanied by a cessation of growth. Other environmental factors, such as drought, induce dormancy in some plants. Along the leafless, winter dormant *stem* (at left) can be seen *foliage leaf scars* that mark the positions of previous seasons' leaves. An *axillary bud* can be seen immediately above each *foliage leaf scar*. These small, latent buds are protected by a covering of tough, scalelike, modified leaves known as bud scales. The *axillary buds* may eventually become active and produce a lateral branch or flowering shoot. Control of their growth is regulated by hormones, and they may or may not become active during the life of the plant. The apple has an alternate leaf arrangement and therefore one *leaf* or one leaf scar per node. Other plant species may have two or more leaves per node. Each *leaf* may have one or more *axillary bud*.

The apex, or tip, of each dormant *stem* is terminated by a *terminal bud*. Like the *axillary bud,* the *terminal bud* has a protective cover of several scalelike, modified leaves called bud scales. In addition, overwintering buds often produce a gummy, resinous substance that provides additional protection from desiccation, disease, and insect infestation. The amount of growth produced during previous seasons can be determined by measuring the distance between sets of terminal *bud scale scars* that mark the position of the *terminal bud* in previous seasons. A single *terminal bud* is produced by each stem tip per season. At the beginning of the growing season, the bud scales of the *terminal bud* are shed, leaving *bud scale scars* to mark their position. Therefore, the distance between these scars is the amount of growth that occurred during one growing season. *Bud scale scars* encircle the *stem* and are bunched because internode elongation does not occur between bud scales.

Color the leaves (F) and flowers (G), as well as the previously listed structures, on the spring shoot.

As spring approaches, bringing longer and warmer days, each stem apex renews hormone production that stimulates growth. The *terminal buds* and *axillary buds* that will begin growth swell as growth of the enclosed stem primordia push outward. As the apex of the *stem* elongates, it pushes out of the bud scales, which soon fall away. The apical meristem at the shoot tips produces leaf primordia that develop into the current season's *leaves*. In the apple, *axillary buds* of the current season located on special lateral *stems,* called fruit spurs, produce *flowers*.

Color the fruit (H), as well as the previously listed structures, on the summer shoot.

Following fertilization in the *flower* in spring, development of *fruit* progresses during summer. As summer passes, the production of hormones subsides and meristematic activity diminishes. The season's *stem* growth is completed and a new *terminal bud* is formed on each *stem* apex. The *leaves* are fully matured, and the developing *fruits* are drawing heavily upon the leave's output of sugars.

Color the fall and winter shoots.

As fall approaches, decreased production of hormones by *stem* tips and *leaves* induces cork cambium formation at the base of each petiole. The layers of dead cork cells in the petiole interrupt the flow of materials between *stem* and *leaves,* resulting in death of the *leaves.* The green chlorophyll pigment breaks down and becomes yellow. This allows other pigments, previously masked by the chlorophyll, to show. The reds, oranges, and yellows of carotenoid pigments and the reds and purples of anthocyanin pigments become evident as fall colors. Soon the dying *leaves* are shed because the cork zone produced in the petiole is a zone of weakness (abscission). In early fall, the production of hormones and ethylene gas by the *fruit* hastens ripening and *fruit* maturation is completed. With the coming of winter, the tree, now enlarged by the current season's growth, once again becomes dormant.

SEASONAL ORGANOGRAPHY.

STEM_A
FOLIAGE LEAF SCAR_B
AXILLARY BUD_C
TERMINAL BUD_D
BUD SCALE SCAR_E
LEAF_F
FLOWER_G
FRUIT_H

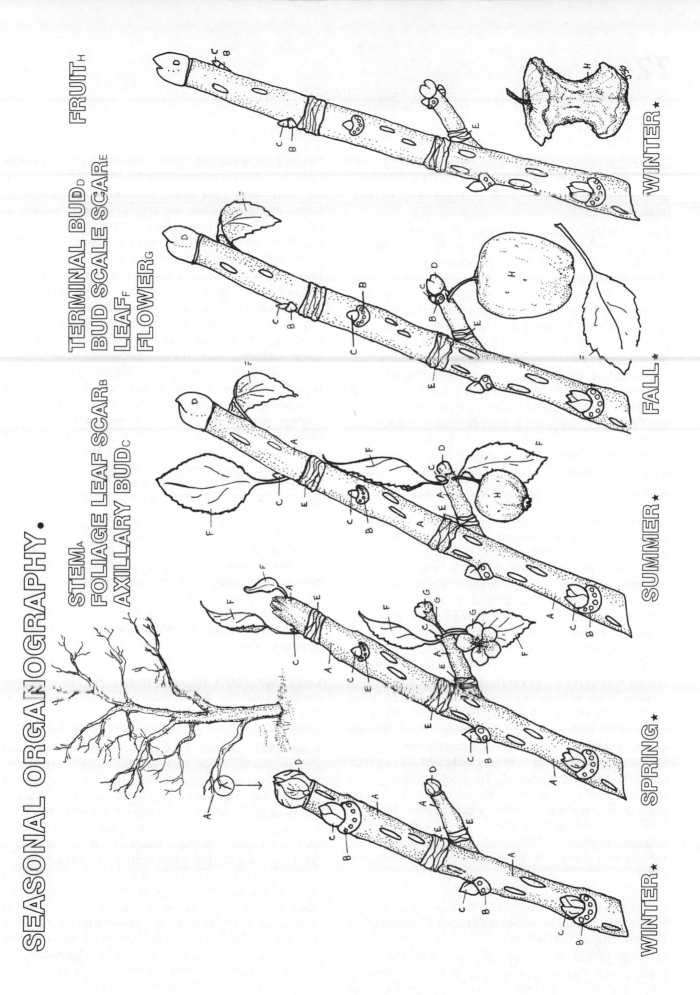

WINTER ★

SPRING ★

SUMMER ★

FALL ★

WINTER ★

72

STORING FOOD RESERVES

Color the roots (A) and stem (B), leaves (C), and renewal buds (D) of the shoot systems on the three illustrations of bulbs at the top of the plate.

Food storage structures primarily function in the storage of excess food materials, mostly in the form of starches and oils, that were produced by photosynthesis during active growth. They may also function in water storage. Food storage structures are most common in herbaceous perennial plants that die back to ground level or below during dormancy. In most of these plants, the food storage structures function as a reserve energy source for the maintenance of life processes during the dormant period when the plant has no photosynthetic surfaces to manufacture food materials. In biennial and perennial plants, the stored food reserves also provide the plant with an adequate energy supply for rapid shoot establishment during the next growing season. In some plants, food storage structures function as asexual reproduction units, with an ample self-contained energy source to facilitate the successful establishment of new plants.

Food storage structures that function to carry a plant through dormancy are usually produced below ground and consist of highly modified *roots, stems,* or *leaves.* The renewal *buds,* the *buds* that will initiate the next season's growth, are usually at or below ground level. On the other hand, food storage structures that double as asexual reproductive units are produced either above or below ground.

True bulbs consist of compact clusters of thick, fleshy, food storage *leaves.* The reduced *stem* is buried within the *leaves* at the base of the bulb. *Roots* seasonally grow from the base of the *stem,* which has one or more renewal *buds.* An onion bulb consists of concentric layers of tubular, fleshy, food storage *leaves.* Each *leaf* is shaped like a narrow-necked pot. Onions produce one to few renewal *buds,* but a garlic bulb typically has several. In a garlic bulb, each renewal *bud,* along with its surrounding *leaves,* usually one to a few, is called a garlic clove. In many bulbous plants, such as many lilies, the fleshy food storage *leaves* are scalelike and overlapping instead of tubular and sheathing, as in the bulbs of onions and garlics.

Color the roots (A), stem (B), leaves (C), and buds (D) of the corm, rhizome, and tuber illustrated in the middle of the plate.

In some plants, the *stems* or portions of the *stems* form dense, fleshy food storage structures. Corms, fleshy rhizomes, and tubers are types of food storage *stems.* Corms are enlarged, vertically oriented *stems* that often have a flattened globose shape. Since they are *stems,* they have nodes and internodes, but little internodal elongation occurs. A corm's surface is often covered with overlapping layers of dry, papery, scalelike *leaves. Roots* may arise from the top of the corm, throughout the corm, or from its base, depending on plant species. The one to few renewal *buds* are located at the top of the *stem.* Corms are typically expended each season, and a new one arises from the top of the previous season's corm, which forms a hardened "basal plate" (not colored) below the new corm. Gladiolas and crocuses produce corms.

Rhizomes and tubers are usually horizontally positioned. Rhizomes are underground *stems* that elongate by seasonal growth. Those modified for food storage, such as an iris rhizome, are thick and fleshy. Growth, due to a terminal renewal *bud* at the tip of each rhizome branch, produces the current season's *leaves* and additional rhizome material. Rhizome branches develop from the axillary *buds* along the rhizome. *Roots* arise along the length of the rhizome.

Tubers are fleshy, enlarged, food storage stem structures produced on a slender, elongate rhizome or on an above-ground *stem.* In the potato, a slender rhizome is terminated by a tuber. The "eye" of a potato tuber (*stem*) consists of a reduced *leaf* (the "eyebrow") and one or more axillary renewal *buds.* Each axillary *bud* may develop into a shoot. In other plants, such as the air potato, above-ground *stems* produce tubers at nodes.

Color the roots (A), stems (B), leaves (C), and buds (D) on the illustrations of storage roots at the bottom of the plate.

Food storage roots, both clustered tuberous roots and some specialized tap roots, are found in many plant groups. Some plants, such as dahlias and buttercups, have tuberous roots; carrots have tap roots for storage. The renewal *buds* of most storage roots are at the base of the *stem* at the top of the *root.* In these, root material alone will usually not produce a new shoot. In other storage roots, such as the sweet potato, renewal *buds* form scattered on the *root.*

As regions of concentrated food storage, many of these structures are used by humans as a food source. Also, since most of these structures are dormant at some time, they provide a convenient unit for storage and transportation. Most have historic and contemporary importance in human mediated dispersal, but due to their often subterranean position, many have limited importance in natural dispersal.

FOOD STORAGE STRUCTURES.

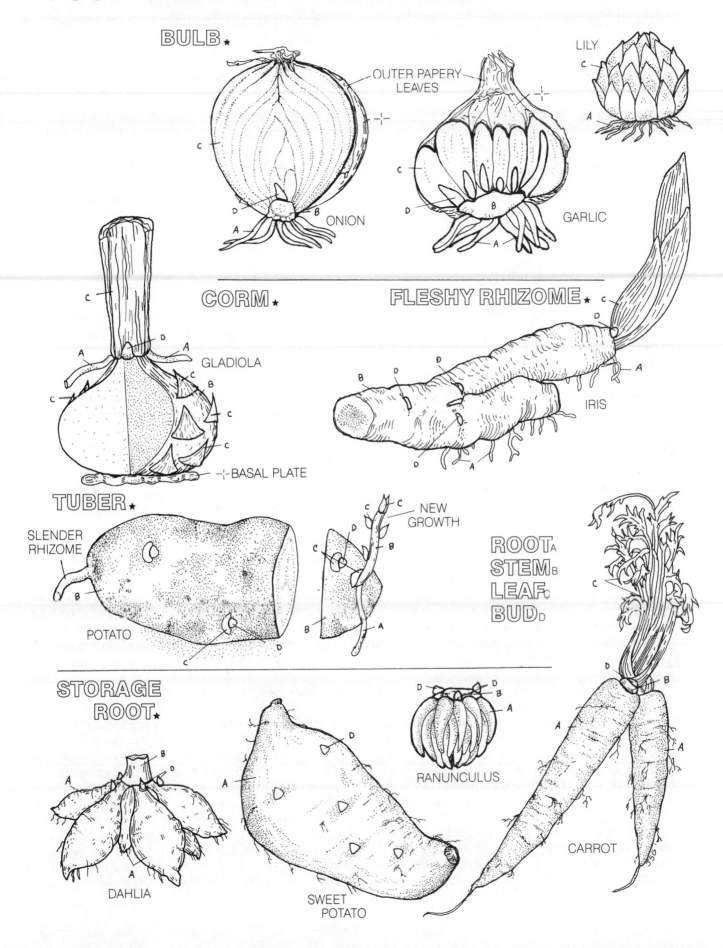

BULB ★

OUTER PAPERY LEAVES

ONION

GARLIC

LILY

CORM ★

GLADIOLA

BASAL PLATE

FLESHY RHIZOME ★

IRIS

TUBER ★

SLENDER RHIZOME

POTATO

NEW GROWTH

ROOT A
STEM B
LEAF C
BUD D

STORAGE ROOT ★

RANUNCULUS

CARROT

DAHLIA

SWEET POTATO

73

NON-SEED-BEARING VASCULAR PLANTS

The ferns and the fern allies contain four divisions of vascular plants.

Color the gametophyte (A), sporophyte (B), symbols for homospory (C), and the border for the Psilophyta (E) at upper left.

The *Psilophyta* contains three species in two genera, *Psilotum* (illustrated) and *Tmesipteris* (not illustrated). *Psilotum* is widely distributed, but localized, in tropical and semitropical regions. *Tmesipteris* is limited to the Australian and New Zealand region. All members are *homosporous,* and the *gametophytes* are subterranean and irregularly branched. Both antheridia and archegonia are produced by a single *gametophyte.* In the *Psilophyta,* the mature *sporophytes,* which lack true roots and true leaves, consist of stems only. Below the surface, the branched stems of *Psilotum* function in support as well as water and nutrient uptake. The above-surface stem is green, photosynthetic, and dichotomously (Y-branched) branched. Though no true leaves are present, small, scalelike microphylls occur in a spiral arrangement along the stem. Meiosporangia occur in fused clusters of three immediately above fertile microphylls.

Color the border and contained diagrams of the Lycophyta (F), including the heterospory (D) symbol.

The *Lycophyta* contains about a thousand species. The three largest and best known genera are *Lycopodium* (illustrated), the club mosses and ground pines; *Selaginella* (illustrated), the spike mosses; and *Isoetes* (not illustrated), the quillworts, which are all aquatic. *Lycopodium* is *homosporous; Selaginella* and *Isoetes* are both *heterosporous.* The *gametophyte* of *Lycopodium* develops exosporically, but the *gametophytes* of both *Selaginella* and *Isoetes* develop endosporically. *Gametophytes* of *Lycopodium* are usually subterranean, but those of *Selaginella* and *Isoetes* develop within the meiospores on the substrate surface.

The *sporophytes* of all *Lycophyta* have true roots, stems, and leaves. In *Lycopodium* and *Selaginella,* the stems are usually dichotomously branched, often prostrate, and thickly covered with numerous scale-like leaves. Meiosporangia are borne at the base of fertile leaves called sporophylls. In some members, the sporophylls are clustered in tight conelike aggregations called strobili.

Color the border and contained diagrams of the Sphenophyta (G).

Sphenophyta (the horsetails) contains only one genus, *Equisetum,* which has about two dozen species. All are *homosporous. Sphenophyta* form pinhead-sized photosynthetic surface-dwelling *gametophytes* that are convoluted and lobed. A single *gametophyte* produces both antheridia and archegonia.

The *sporophytes* of all species are somewhat similar vegetatively. In the *Sphenophyta,* the *sporophytes* have a branched underground rhizome from which upright branches arise. They have true roots, stems, and leaves. The stems are characterized by whorls of small, scalelike leaves about the nodes and may be branched or unbranched.

Four meiosporangia are present on each highly specialized meiosporophyll, which are clustered to form a pronounced strobilus.

Color the border and contained diagrams of the Pterophyta (H).

The *Pterophyta,* the true ferns, is the largest division, containing about twelve thousand species distributed worldwide. Except for a small group of *heterosporous* water ferns, the ferns are *homosporous. Gametophytes* in many are heart-shaped, thumbnail-sized, surface-dwelling, and one to a few cells thick. Rhizoids, present on the undersurface, function in anchorage and water and mineral nutrient uptake. Both antheridia and archegonia are usually produced on a *gametophyte.* The *sporophytes* range from small plants a few centimeters tall to large, treelike plants to about 20 meters, but most produce creeping rhizomes. As expected due to their presence in a broad range of environments and habitats, the *sporophytes* exhibit much morphological diversity. All have true roots, stems, and leaves (fronds). Meiosporangia are usually produced in clusters on the underside of fertile fronds, which, depending on species, may or may not resemble the vegetative fronds.

FERNS AND FERN ALLIES.

GAMETOPHYTE_A   SPOROPHYTE_B

LYCOPHYTA_F

PSILOPHYTA_E

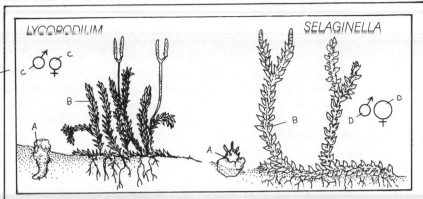

LYCOPODIUM   SELAGINELLA

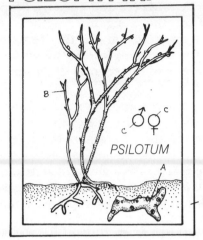

PSILOTUM

E⎯ HOMOSPOROUS_C   SPHENOPHYTA_G

HETEROSPOROUS_D

PTEROPHYTA_H

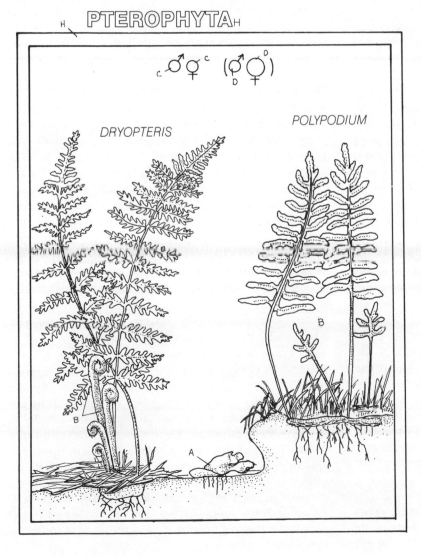

DRYOPTERIS   POLYPODIUM

EQUISETUM

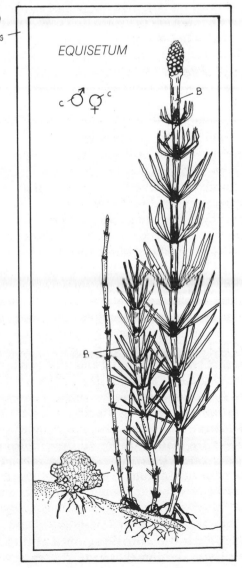

74
SELAGINELLA (LYCOPHYTA)

The spike mosses, genus *Selaginella,* are a small group, less than a thousand species, of mostly tropical fern allies that closely resemble club mosses, genus *Lycopodium.* Unlike the other fern allies and all but a very few ferns, the spike mosses are heterosporous rather than homosporous. A general spike moss life history provides a good example of the life history of a heterosporous plant.

Starting at the upper left and continuing down and across, color the sporophyte (A), strobilus (A¹), microsporophyll (A²), microsporangium (B), microspores (C), megasporophyll (A³), megasporangium (D), and megaspores (E). Stop coloring just before the mitosis symbols at the center of the plate.

At the tips of some branches of the *sporophyte,* aggregations of sporangium-bearing leaves, called sporophylls, form loose, conelike structures called *strobili.* Each sporophyll bears one sporangium (meiosporangium) near its base. Each *strobilus* usually contains two kinds of sporophylls, *microsporophylls* bearing *microsporangia* and *megasporophylls* bearing *megasporangia.* The outer wall of the sporangia is formed by a sterile jacket of cells that surrounds and protects the enclosed sporogenous tissue, the diploid sporocyte cells that will undergo meiosis to form meiospores. In the *microsporangium,* several mitotic divisions produce numerous microsporocytes (not shown), which are the diploid cells that directly undergo meiosis to form numerous haploid *microspores.* In the *megasporangium,* no mitotic divisions occur; so the entire contents of the *megasporangium* forms one large megasporocyte (not separately colored) that, by meiosis, produces four haploid *megaspores.* Because of this difference in development, there is a large size difference between the *microspores* (meaning "small spores") and the *megaspores* (meaning "large spores"). Upon rupture of mature sporangia, the small, lightweight *microspores* are readily dispersed by air currents; but the larger, heavier *megaspores* often remain in place in the *strobili.* Copious amounts of *microspores* are carried by the wind, and some eventually contact *strobili* and sift down to the exposed *megaspores.*

Starting at the mitosis symbol at the center, color the male gametophyte (F), microspore wall (C¹), sperm (G), female gametophyte (H), mega- spore wall (E¹) archegonia (I), and eggs (J). Stop coloring just before the syngamy symbol at lower left. The first diagram of the female gametophyte is a top surface view of the whole female gametophyte, and the second diagram is a sectioned side view.

Germination of a *microspore* produces a *male gametophyte* (microgametophyte) that develops totally within the confines of the *microspore wall* (endosporic). At maturity, the entire *male gametophyte* consists of a single-cell-layer-thick antheridial jacket, or *antheridium,* and fertile tissue that forms biflagellated *sperm. Sperm* are released by rupture of the *microspore wall.*

Megaspores germinate to begin development of a *female gametophyte* (megagametophyte) within the confines of the *megaspore wall* (endosporic), but as the *female gametophyte* enlarges, the *megaspore wall* splits along three sutures to expose the upper surface of the *female gametophyte.* However, the *female gametophyte* remains mostly contained within the *megaspore wall.* Absorptive hairlike structures, rhizoids (not separately colored), which absorb water and dissolved minerals, develop on the *female gametophyte* at the ends of the splits in the *megaspore wall. Archegonia* also develop on the exposed upper surface of the *female gametophyte.* Each *archegonium* contains a single *egg,* and at maturity, the neck and ventral canal cells (not shown) disintegrate to form an open passageway to the *egg. Sperm,* attracted by substances released from the matured *archegonium,* swim through a film of water and enter the open passageway.

Color the previously listed features and the zygote (K) on the remaining diagrams.

One *sperm* eventually fertilizes the *egg* to form a *zygote.* Development of the embryonic *sporophyte* occurs within the *archegonium* in the *female gametophyte.* Though a number of *archegonia* are produced by each female gametophyte, usually a single embryo develops. During its early development as an embryo, the *sporophyte* is parasitic upon the *female gametophyte,* from which it obtains all its requirements. Breakdown of the parent sporophytic *strobili* releases the germinated *megaspores (female gametophyte),* and the young *sporophytes* soon develop a root and shoot system to become independent plants.

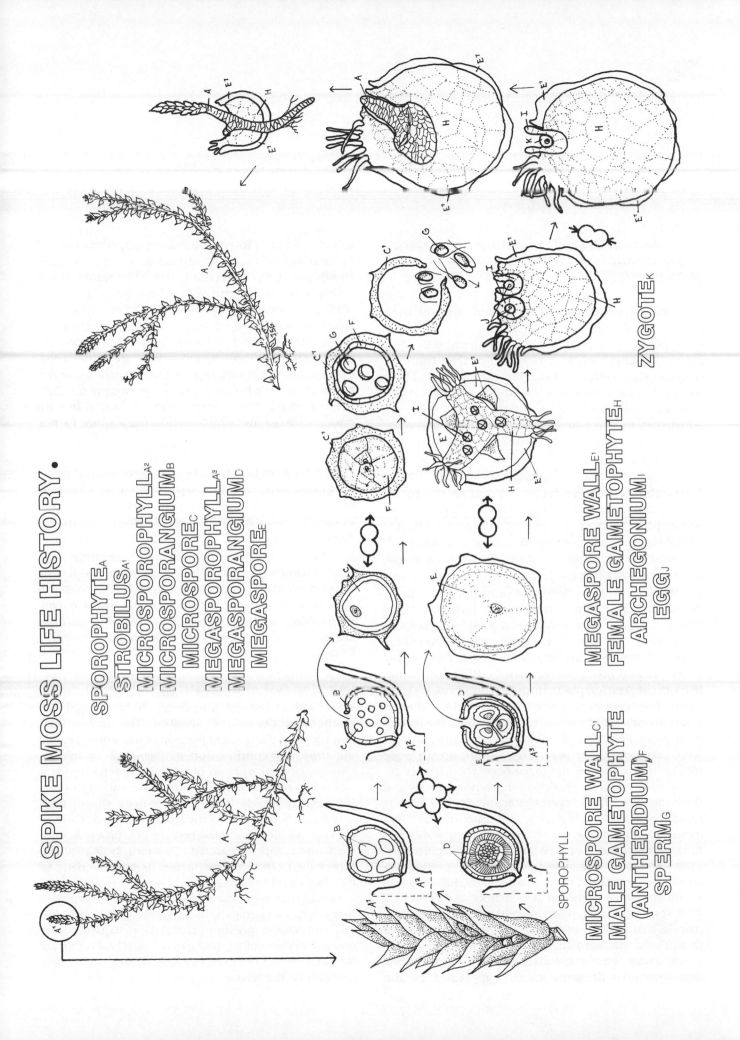

SPIKE MOSS LIFE HISTORY.

SPOROPHYTE$_A$
STROBILUS$_{A^1}$
MICROSPOROPHYLL$_{A^2}$
MICROSPORANGIUM$_B$
MICROSPORE$_C$
MEGASPOROPHYLL$_{A^3}$
MEGASPORANGIUM$_D$
MEGASPORE$_E$

MICROSPORE WALL$_{C^1}$
MALE GAMETOPHYTE
(ANTHERIDIUM)$_F$
SPERM$_G$

MEGASPORE WALL$_{E^1}$
FEMALE GAMETOPHYTE$_H$
ARCHEGONIUM$_I$
EGG$_J$

ZYGOTE$_K$

75
TRUE FERNS (PTEROPHYTA)

Color the roots (A) and stems (B) of the two large, diagrammatic ferns at the top and bottom of the plate.

This plate provides a brief introduction to the morphology of the fern sporophyte.

Ferns typically produce most of their *roots* from along the length of their *stem* rather than from the original root system formed by the embryo. These *stem*-originated *roots* are called adventitious *roots*. Most ferns do not have an upright growth form; instead, they produce *stems,* called *rhizomes,* that grow closely appressed to the substrate surface or just beneath it.

Color the vegetative frond (C) and fertile frond (D) on the right half of the large, diagrammatic, dimorphic fern at the top of the plate. Do not color the other fronds on this fern yet. Also color the two examples of dimorphic ferns at the center of the plate.

Perhaps the greatest degree of morphological diversity is present in the leaves that in ferns are called fronds. Some ferns produce dimorphic (meaning "two forms") fronds. That is, two rather different-appearing frond types are produced. In many dimorphic ferns, the *vegetative* (sterile) fronds differ significantly from the *fertile* (sporangium-bearing) fronds. Other dimorphic ferns produce two different kinds of *vegetative* fronds that function to better adapt the fern to its habitat. For example, *Salvinia,* a floating aquatic fern, produces three *vegetative* fronds at each node on the stem. Two of the three *vegetative* fronds are broad and entire, and function in photosynthesis and flotation. The third *vegetative* frond at each node is submerged and highly dissected into narrow segments and functions in the absorption of water and mineral nutrients. Another fern with dimorphic *vegetative* fronds is the staghorn fern, *Platycerium,* an epiphytic (meaning "upon plant") fern that grows as an independent plant on tree trunks and branches. This fern has overlapping, shield-shaped *vegetative* fronds that grow appressed to the substrate (tree trunk) and staghorn-shaped fronds that grow outward, away from the substrate. The shield-shaped *vegetative* fronds grow closely appressed to the

substrate in their lower half but often spread outward in their upper half. The funnellike aspect of these fronds serves to catch falling debris and water. As the debris decomposes, the released nutrients, as well as water, are absorbed by *roots* that grow among the layers of shield-shaped *vegetative* fronds. Only the youngest of these fronds are green and photosynthetic. The staghorn-shaped fronds are either *vegetative* fronds that function in photosynthesis or *fertile* fronds with broad areas of *sporangia* on their lower surface. These *fertile* fronds function in both photosynthesis and reproduction (meiospore formation).

Color the previously listed features on the large diagrammatic, monomorphic fern at the bottom, but do not color the other fronds yet. Also color the monomorphic fern example at bottom right.

In ferns that produce monomorphic (meaning "one form") fronds, all fronds, both *vegetative* and *fertile,* are very similar in appearance. Boston fern *(Nephrolepis exaltata),* rabbit's foot fern *(Davallia trichomanioides),* and sword fern *(Polystichum munitum)* are all examples of ferns that produce monomorphic fronds.

Color the rest of the plate.

The frond, or leaf, of most ferns can be divided into morphologically distinct sections. The narrow stalk that supports the laminar portion of the frond is called the *stipe;* the continuation of this stalk, or midrib, into the laminar portion of compound fern fronds is called the *rachis;* and the leaflets are called *pinnae.* Developing fronds have a fiddle-neck shape. Each new frond is coiled with the frond tip at the center of the coil. As the frond matures, the coil unwinds until the frond is fully expanded. This morphological feature, called *circinate vernation,* is characteristic of fern leaf development.

Fertile fronds produce areas densely covered with *sporangia* or aggregations of *sporangia* called sori. The size, shape, position, maturation sequence, and other morphological features of sori are constant within a species but exhibit a high degree of diversity throughout the ferns.

FERN FEATURES.

SPOROPHYTE★
  ROOTᴀ
SHOOT★
    RHIZOME (STEM)ʙ
  FROND★
    VEGETATIVEᴄ
    FERTILEᴅ
    STIPEᴇ
    RACHISꜰ
    PINNAɢ
    CIRCINATE
      VERNATIONʜ
SPORANGIAı

DIMORPHIC FRONDS ★

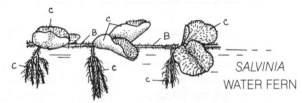

SALVINIA
WATER FERN

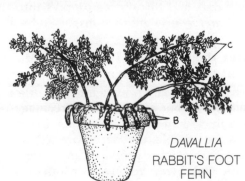

PLATYCERIUM
STAGHORN FERN

MONOMORPHIC FRONDS ★

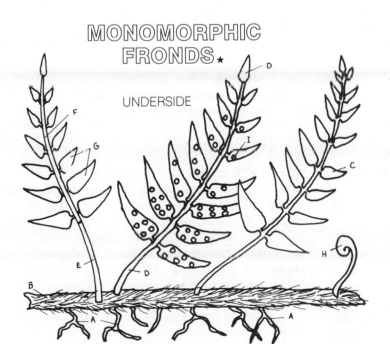

UNDERSIDE

DAVALLIA
RABBIT'S FOOT FERN

76

TRUE FERNS (PTEROPHYTA)

Begin coloring with the sporophyte (A) at the upper left and continue clockwise. Stop with the diagram following the meiosis symbol.

The fern generation commonly grown as an ornamental plant is the *sporophyte*. Environmental factors, such as day length and temperature, influence the seasonal development of fertile fronds on mature *sporophytes*. Fertile fronds may appear similar to vegetative fronds or may be quite different, or dimorphic, in appearance. The presence of *sori* on the undersides of fertile fronds distinguishes them from vegetative fronds. Each *sorus* consists of a cluster of numerous stalked *sporangia*. A protective cover, the *indusium,* over each cluster of *sporangia* may or may not be present in the *sorus*, depending on species. Within each *sporangium* are many diploid spore mother cells called sporocytes (not shown). The *annulus,* a structure composed of a series of thick-walled cells along the crest of the *sporangium,* is part of the sporangial wall.

As the *sporangium* matures, the enclosed spore mother cells undergo meiosis to produce the first cells of the haploid generation, the *meiospores*. In most ferns, *meiospore* release occurs over an extended period of time since all *sporangia* within a *sorus* do not mature at the same time. As the *sporangia* in a *sorus* mature, the *indusium* shrivels so that *meiospore* release is unimpeded. *Sporangium* opening for *meiospore* release is controlled by the *annulus*. As the *sporangium* drys at maturity, the cells of the *annulus* shrink and place tension on the *sporangium* wall. When sufficient tension develops, the *sporangium* bursts and throws *meiospores* into the air. The freed *meiospores* are then dispersed by air currents.

Color the structures (F) through (M). Stop coloring at the syngamy symbol.

Ferns produce and release into the air a multitude of *meiospores,* but only a small percentage land in sites favorable for germination. If environmental conditions are suitable, the *meiospore* will split open, and mitotic divisions produce an independent-living, multicellular *gametophyte* or prothallus. Fern *gametophytes* are usually small, inconspicuous, heart-shaped flaps of haploid tissue, one to a few cells thick, that grow on moist surfaces. *Rhizoids,* small filamentous structures, anchor the *gametophyte* in the

substrate and absorb water and mineral nutrients. The *gametophyte* develops outside the *meiospore* wall (exosporic) and produces gamete-forming structures, gametangia, at maturity. Most ferns are homosporous. That is, the *meiospores* are all alike in size and shape. In most ferns, a single *gametophyte* will produce both male *(antheridia)* and female *(archegonia)* gametangia. The *antheridia* usually mature before the *archegonia* on the same *gametophyte*. *Antheridia* are usually located among the *rhizoids* on the pointed end; *archegonia* are found clustered in the notch area of the heart-shaped *gametophyte*. Even though male and female gametangia are adjacent to one another, self-fertilization is discouraged by their sequential maturation.

When an *antheridium* is mature and sufficient water for *sperm* movement is present, the *antheridium* cover cell opens and the motile *sperm* swim out. Under similar conditions, the cover cells at the tip of the *archegonium* neck open. The *neck* and *ventral canal cells* degenerate, leaving an open passageway, the neck canal, to the *egg*. Chemical substances that serve as *sperm* attractants are subsequently released by the *archegonium*.

Color the rest of the plate.

Guided by the attractants, *sperm* seek out the open *archegonium* and enter the neck canal. Syngamy occurs when one *sperm* fuses with the *egg* to form the first cell of the diploid generation, the *zygote*. Early *sporophyte* development takes place within the archegonial chamber, which enlarges to accommodate the growing embryo. The young *sporophyte* begins its life as a dependent, parasitic plant that receives all nutritional needs and water from the *gametophyte*. A mass of sporophytic tissue, the foot, forms within the archegonial chamber during early development and functions in the absorption of nutrients from the *gametophyte*. As development of the *sporophyte* progresses, a shoot and root system capable of supporting the *sporophyte* is produced. While still quite small, the *sporophyte* becomes an independent-living plant. The spent *gametophyte* soon dies and decomposes. The nutritional demands of the young, parasitic *sporophtye* on the *gametophyte* usually lead to the formation of only one *sporophyte* per *gametophyte* even though many *archegonia* are present.

FERN LIFE HISTORY.

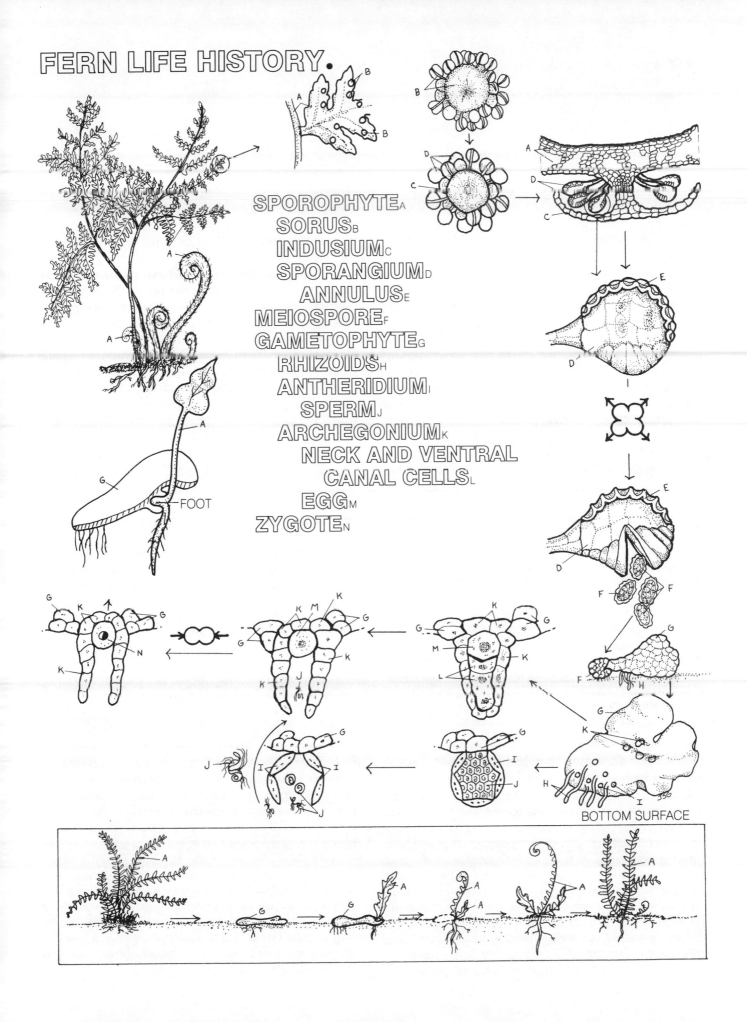

SPOROPHYTE_A
SORUS_B
INDUSIUM_C
SPORANGIUM_D
ANNULUS_E
MEIOSPORE_F
GAMETOPHYTE_G
RHIZOIDS_H
ANTHERIDIUM_I
SPERM_J
ARCHEGONIUM_K
NECK AND VENTRAL
CANAL CELLS_L
EGG_M
ZYGOTE_N

FOOT

BOTTOM SURFACE

77

PROTECTION OF THE EGG

Color the diagrams labeled "archegonium" across the top of the plate.

A major adaptation of the *gametophyte* of true land plants is the evolution of the *archegonium.* During development, the *egg* is completely surrounded by the *archegonium* wall. Four specialized cells, the *cover cells,* close over the tip of the *archegonium,* and numerous cells form the narrow *neck* and swollen base or *venter.* One binucleate *neck canal cell* to several *neck canal cells* and a single *ventral canal cell* (immediately above the *egg*) block the neck canal during *egg* development. When the *egg* is mature and receptive to sperm, the *cover cells* open outward and the *neck canal cells* and *ventral canal cell* break down to create an open passageway to the *egg.* This permits the entry of sperm into the *venter,* where the *egg* is located. One sperm fertilizes the *egg* to form a *zygote,* which is retained within the *archegonium* during early *sporophyte* development.

During its early development, the *sporophyte* depends completely on the *gametophyte* for nutrients and water, which pass from the vegetative portion of the *gametophyte* through the *archegonium* to the young *sporophyte,* thus allowing for sheltered establishment of the *sporophyte.* In all land plants except bryophytes, the *sporophyte* outgrows the *archegonium* and becomes an independent plant. Though structural features differ, *archegonia* are formed by the *gametophyte* of all true land plants, including flowering plants.

Color the diagrams labeled "exposed megasporangium."

All the more specialized land plants, including all gymnosperms and flowering plants, are heterosporous. *Sporophytes* of heterosporous plants form two kinds of meiosporangia, microsporangia that produce microspores (not shown) and *megasporangia* that produce *megaspores.* The *megaspores* of heterosporous plants are heavy and not readily dispersed. One significant advantage they have over homosporous meiospores is the large amounts of reserve food materials within the *megaspore,* which permit early endosporic (within the megasporic wall) development of the female *gametophyte.* This adaptation in-creases the potential for successful establishment of the female *gametophyte.*

In plants that form exposed *megasporangia,* the *megaspores* are released by rupture of the *megasporangium.* If deposited in a favorable habitat, the *megaspores* take in water and germinate within the *megaspore wall.* As the *megaspores* grow, they split the *megaspore wall,* and a portion of the female *gametophyte* becomes exposed. *Archegonia* usually form on the exposed surface of the female *gametophyte.* Fertilization of an *egg* within an *archegonium* forms a *zygote* that undergoes meiosis to form an embryonic *sporophyte.* Early development of the *sporophyte* is within the *archegonium* that is imbedded in the endosporic female *gametophyte.*

Color the rest of the plate.

One major adaptation of seed plants is the presence of a multicellular layer of *sporophyte* tissue, called the *integument,* that is fused to the *megasporangium.* The *integument* completely encloses the *megasporangium* except for a small pore, the *micropyle.* This integumented *megasporangium,* called an ovule, is the characteristic feature of seed plants. Each *megasporangium,* called the *nucellus* in seed plants, contains one (illustrated) or more megasporocytes that, through meiosis, each produce four *megaspores.* In most seed plants, all but one of the four *megaspores* from each megasporocyte abort and are reabsorbed.

Deep within the ovule, each functional *megaspore* forms a female *gametophyte* that produces one or more *archegonia.* In seed plants, the *archegonia* may be highly reduced. Since the female *gametophyte* develops imbedded within *sporophyte* tissues (the *integument* and the *nucellus*), the *gametophyte* depends completely upon the *sporophyte.* Because of these surrounding tissues, fertilization of an *egg* requires adaptations that permit the transfer of sperm through these tissues to the *egg* (not shown). Once fertilized, an *egg* becomes a *zygote* that undergoes mitosis to begin embryonic *sporophyte* development. After a period of initial development, the *sporophyte* becomes dormant and the *integument* becomes hardened (sclerified) to form a *seed coat.* Food reserve materials are stored as female *gametophyte* tissue in gymnosperms (shown).

REPRODUCTIVE SPECIALIZATION.

GAMETOPHYTIC.
ARCHEGONIUM ★

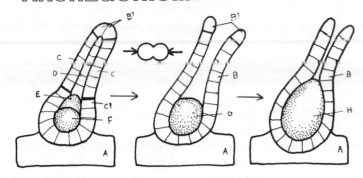

GAMETOPHYTE A
  ARCHEGONIUM B
    COVER CELLS B1
    NECK C
    VENTER CELLS C1
    NECK CANAL CELLS D
    VENTRAL CANAL CELL E
    EGG F
ZYGOTE G
SPOROPHYTE H
  MEGASPORANGIUM
    (NUCELLUS) I
  MEGASPORE J
  MEGASPORE WALL J1
INTEGUMENT
  (SEED COAT) K
MICROPYLE L ¦

SPOROPHYTIC.
EXPOSED
MEGASPORANGIUM ★

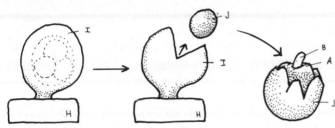

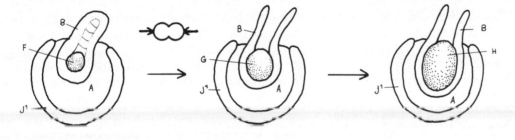

INTEGUMENTED MEGASPORANGIUM ★

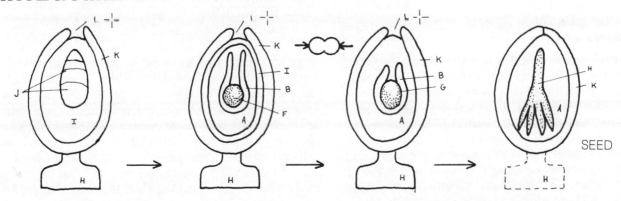

SEED

78

SEED-BEARING VASCULAR PLANTS

Gymnosperms contain four divisions of seed plants. Included in the gymnosperms are the divisions *Cycadophyta,* the cycads with about eighty species; *Ginkgophyta,* with only one species *(Ginkgo biloba); Gnetophyta,* a diverse assemblage of about forty species; and *Coniferophyta,* the cone-bearing plants with about five hundred species.

Color the female (A) and male (B) reproductive structures and the cycad plant representing the Cycadophyta (C), as well as the surrounding border.

The *Cycadophyta,* or cycads, are distributed primarily in tropical and subtropical areas. Sporophytes have columnar, upright, and usually unbranched stems from a few centimeters to about twenty meters in length, but most are less than three meters. The stems bear numerous, large, usually pinnately, compound leaves in a crownlike cluster at their tips. Because of this, most cycads superficially resemble tree ferns or palm trees. The portion of the stem below the crown of leaves is naked, without leaves, since leaves are shed within a few years. Leaf abscission usually occurs up to a few centimeters out on the petiole; so old leaf bases sheath the stem. In most species, the microsporangia are borne on microsporophylls aggregated into a conelike *male* strobilus, and the ovules are produced within conelike *female* strobili. *Female* strobili are usually larger than *male* strobili. All species produce *male* strobili and *female* strobili on separate sporophytes.

Color the border and enclosed diagrams of the Ginkgophyta (D).

The *Ginkgophyta* contains a single species, *Ginkgo biloba,* that occurs naturally only in a small area in southern China. However, its popularity as an ornamental plant has given it widespread distribution. *Ginkgo biloba* is a highly branched deciduous tree, to more than twenty-five meters high, that has pale green, fan-shaped leaves that turn bright yellow in autumn. Two types of shoots, both of which bear a single leaf at each node, are produced. The long shoots exhibit internodal elongation and account for most seasonal stem growth; the short shoots, called spur shoots, exhibit little or no internodal elongation

and bear the reproductive structures and most of the leaves. The microsporangia are aggregated in elongate, pendant *male* strobili, and the ovules are borne in pairs (a highly reduced *female* strobilus) on a stem. As in the cycads, *male* and *female* reproductive structures occur on separate sporophytes.

Color the border and enclosed diagrams of the Gnetophyta (E).

The *Gnetophyta* contains three·dissimilar genera. *Gnetum* (not illustrated) is a highly branched tropical shrub with broad, laminar leaves. *Welwitschia* (not illustrated), from Western Africa, has a short, broad stem with only two strap-shaped, vegetative leaves that have basal meristems that continue to produce new leaf tissue. Thus, the leaves continue to increase in length as long as the plant lives, but the tips become frayed and abraded. *Ephedra,* or Mormon tea, of southwestern North America, is the largest and most widespread genus.

Ephedra is a shrubby plant with highly branched stems that bear clusters of two to three small, scalelike leaves at each node. The microsporangia are aggregated into short, erect *male* strobili; the ovules are borne singly within each short, erect *female* strobilus. *Male* strobili and *female* strobili, on separate sporophytes in most species, occur in clusters at the nodes.

Color the border and enclosed diagrams of the Coniferophyta (F).

The *Coniferophyta,* or cone-bearing plants, are primarily distributed in the cooler temperate zones. All are woody perennials and almost all are trees. The leaves are usually small and scalelike, as in junipers, or large and needlelike, as in pines. Some, such as *Podocarpus,* produce flattened, lance-shaped leaves. Most shed the leaves after a few to several years, but a few species are deciduous. The microsporangia, typically aggregated in small *male* cones produced at or near the tips of branches, consist of few to numerous microsporophylls. The larger, ovule-bearing *female* cones are usually woody at maturity and are much more structurally complex than the *male* cones. *Female* cones are also produced at or near branch tips. Most species produce both *male* and *female* cones on the same sporophyte.

GYMNOSPERMS.

REPRODUCTIVE STRUCTURES. ★
FEMALEA
MALEB

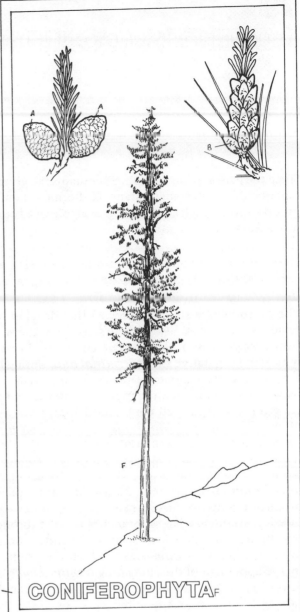

CONIFEROPHYTAF

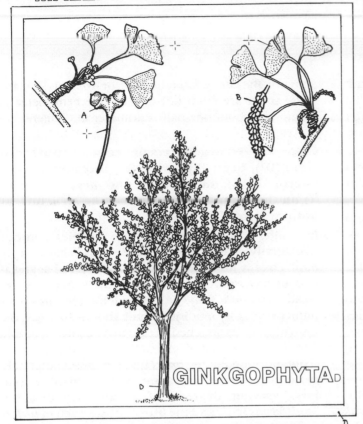

GINKGOPHYTAD

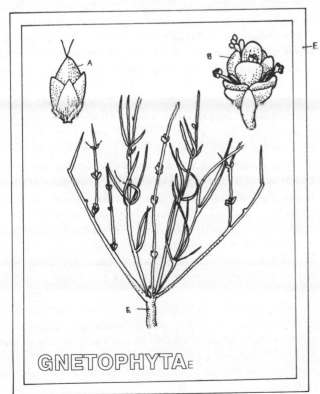

GNETOPHYTAE

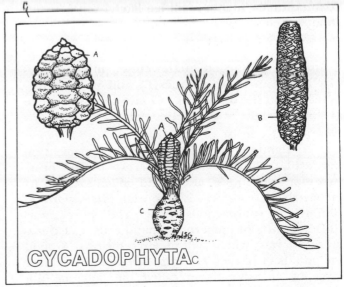

CYCADOPHYTAC

79
CONIFERS (CONIFEROPHYTA)

Since the genus *Pinus,* the pines, contains the greatest number of conifer species and is the most familiar group of conifers, it is used as an example of a conifer life history.

Color the male cones (A), female cones (D), and terminal buds (J) on the three habit diagrams at the left.

The reproductive shoot primordia that develop into either *male cones,* also called staminate cones, or *female cones,* also called ovulate cones, are initiated within the vegetative shoot *terminal buds* during late spring or early summer. During summer, fall, and winter of the first year, the minute primordia of both cone types remain well concealed within the tissues of the dormant *terminal buds.* When the *terminal buds* begin active growth in late winter or early spring during the first year, the developing *male cones* and *female cones* become visible along the elongating *terminal buds* (as illustrated). *Male cones* are typically produced in dense clusters near the base of *terminal buds* on secondary lateral branches located throughout the tree, but they are especially abundant on the lower branches. *Female cones,* which are produced on the upper half of the elongating *terminal buds,* are fewer in number and are located primarily at the tips of major branches in the upper portions of the tree. This positioning enhances cross-pollination. Both cone types are highly modified lateral reproductive shoot systems.

Color the enlarged male cone, microsporophylls (A¹), microsporangia (B), and microsporocytes (C) on the diagrams of male cone structure at the top right of the plate.

Male cones consist of numerous *microsporophylls* that are tightly clustered in a spiral arrangement on the fertile shoot axis. Each *microsporophyll* bears two *microsporangia,* also called pollen sacs, on its lower (abaxial) surface. Within each *microsporangium* is the sporogenous tissue. The sporogenous tissue consists of numerous diploid cells, called *microsporocytes,* which will undergo meiosis. Around the periphery of each *microsporangium* is a cellular layer of nutritive tissue called the tapetum (not

shown). By late winter (just prior to meiosis), the *male cones* are from one-half to one centimeter in length and about one-half centimeter in diameter.

Color the enlarged female cone, ovuliferous scale (D¹), bract (E), ovule (F), integument (F¹), micropyle (G), nucellus (H), and megasporocyte (I) on the diagrams of female cone structure at bottom right.

In conifers, the *female cones* are complex and not structurally comparable to *male cones.* That is, a *female cone* is not simply a tightly clustered aggregation of megasporophylls on a fertile shoot axis. Instead, a *female cone* is formed by the fusion of numerous highly modified fertile shoots. In pines, the individual units attached to the single central axis of a *female cone* consist of an *ovuliferous* (ovule-bearing) *scale* and a subtending *bract* that is almost completely fused to the *ovuliferous scale* above it. (In other conifers, varying degrees of fusion between the *ovuliferous scale* and the *bract* occur). Each *ovuliferous scale* is formed by the fusion of megasporophylls (not shown) and other fertile shoot components.

On the upper (adaxial) surface of each *ovuliferous scale* are two *ovules.* The *ovules* are oriented with their *micropyles* toward the central cone axis and are partially imbedded in the tissues of the *ovuliferous scale.* Each *ovule* has an *integument* (one multicellular layer) that, except for the *micropyle* completely surrounds the megasporangium, which is also called the *nucellus.* In pines, each *nucellus* contains a single *megasporocyte.* The ample *nucellus* functions as a nutritive tissue, and the *megasporocyte* is the diploid cell that will undergo meiosis. A small chamber, called the micropylar chamber (not separately colored), is located within each *ovule* between the *nucellus* and the *micropyle.* By late winter, just prior to meiosis, the *female cones* are from one-half to about two centimeters long and about one centimeter in diameter.

Thus, in pines, *male* and *female cones* are initiated in the summer or fall of one calendar year, and then become exposed and visible in the early spring of the following calendar year, at which time meiosis of the *megasporocytes* and pollination occur almost simultaneously.

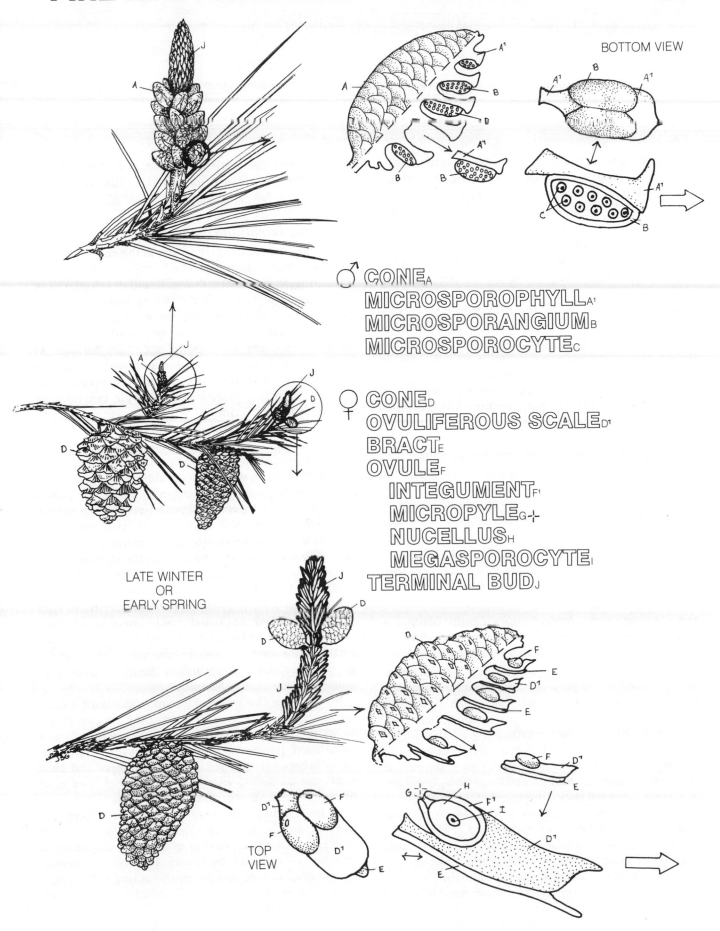

PINE LIFE HISTORY I.

BOTTOM VIEW

♂ CONE_A
MICROSPOROPHYLL_{A¹}
MICROSPORANGIUM_B
MICROSPOROCYTE_C

♀ CONE_D
OVULIFEROUS SCALE_{D¹}
BRACT_E
OVULE_F
   INTEGUMENT_{F¹}
   MICROPYLE_G+
   NUCELLUS_H
   MEGASPOROCYTE_I
TERMINAL BUD_J

LATE WINTER
OR
EARLY SPRING

TOP
VIEW

Color the diagrams of pollen development and release at upper left. Stop coloring when you finish with the enlarged pollen grain.

In the late winter or early spring, a few to several months following male cone initiation, the diploid *microsporocytes* undergo meiosis to form four haploid *microspores*. While still within the *microsporangia,* the *microspores* undergo mitosis to begin *male gametophyte* development endosporically, that is, within the confines of the *microspore wall.* Two mitotic divisions produce a four-celled *male gametophyte.* The combination of *microspore wall* and contained young *male gametophyte* is called a *pollen grain.* As an adaptation for wind-mediated pollen dispersal, two winglike air bladders that increase the surface area to weight ratio to increase buoyancy develop from the *microspore wall.* Meiosis and the two mitotic divisions that form the young *male gametophyte* take place over a short span of time, and in early spring, the *pollen grains* are released by rupture of the *microsporangium.* The multitude of lightweight *pollen grains* are nonselectively dispersed by air currents and are carried only by chance to a female cone.

Color the first two diagrams of ovule development at the bottom of the plate. Continue coloring the pollen grains.

Meiosis within the female cones occurs about the same time as meiosis within the male cones. Following meiosis, the ovules consist of an *integument,* the *nucellus,* and a linear chain of four *megaspores.* Three of the four *megaspores* soon abort and are reabsorbed, leaving one functional *megaspore* within the *nucellus.* In early spring, the axis of the female cone elongates, thereby slightly separating each *ovuliferous scale* from the *bract* above to permit entry of *pollen grains* into the area of the ovules. Within the micropylar chamber and protruding through the micropyle is a drop of sticky fluid, called the *pollination droplet,* that captures *pollen grains.* Soon the *pollination droplet* becomes reabsorbed, carrying the captured *pollen grains* through the micropyle into the micropylar chamber, where they come in direct contact with the *nucellus.* At about the same time, mitotic divisions of the functional *megaspore* have formed a young *female gametophyte,* containing thirty-two free nuclei and no cell walls, within the *nucellus.*

Development of both the *male gametophyte* and *female gametophyte* is almost at a standstill during the summer, fall, and winter.

Color the remaining diagrams.

In the late winter or early spring at the end of the second year of development, the female cones are about four to eight centimeters long. Rapid development of the *male gametophyte* and *female gametophyte* now begins. The pollen tubes of the *male gametophytes* grow through the *nucellus* and into the *female gametophyte,* and two sperm nuclei are formed. At about the same time, the *female gametophyte* undergoes numerous mitotic divisions and forms cell walls. Within the cellular *female gametophyte,* two or more *archegonia,* each containing a single *egg,* are formed near the micropylar end. Through the breakdown and utilization of nutrients from the *nucellus,* the *female gametophyte* grows to occupy most of the ovule.

At fertilization, which occurs about thirteen months following *megaspore* formation in pines, the pollen tube of the *male gametophyte* enters the *egg* within the *archegonium* and discharges its two sperm nuclei, as well as other protoplasmic materials, into the *egg.* One sperm nucleus fuses with the *egg* nucleus to form a zygote, while the other breaks down and is reabsorbed. Since two or more *archegonia* are present within each *female gametophyte,* two or more *eggs* may be fertilized, but usually only one develops into an *embryo.* Development of the *embryo* is rapid through late spring and early summer. As the *embryo* grows, the *female gametophyte* continues to enlarge and utilize tissues of the *nucellus.*

By the time the *embryo* becomes dormant within the seed (the matured ovule containing a dormant *embryo*), the *nucellus* is exhausted and present only as a thin, dried, papery membrane, called the perisperm, located between the *female gametophyte* and *seed coat* (the hardened *integument*). During *seed coat* formation, the micropyle becomes closed.

In most pines, the seeds are released from the female cones between late summer and winter. Each seed has a winglike portion of the *ovuliferous scale* attached to increase buoyancy for wind dispersal. Upon germination, usually in the spring following a required period of cold, a pine *seedling,* with several cotyledons, develops.

PINE LIFE HISTORY II.

MICROSPOROPHYLL_A → MICROSPOROPHYLL A
MICROSPORANGIUM B
MICROSPOROCYTE C
MICROSPORE D
MICROSPORE WALL D'
MALE GAMETOPHYTE E
POLLEN GRAIN F

OVULE ★
  INTEGUMENT G
  MICROPYLE H-¦-
  NUCELLUS I
  MEGASPORE J
OVULIFEROUS SCALE K
BRACT L
POLLINATION DROPLET M
FEMALE GAMETOPHYTE N
  ARCHEGONIUM O
  EGG P
EMBRYO Q
SEED COAT G'
  WING K'
SEEDLING R

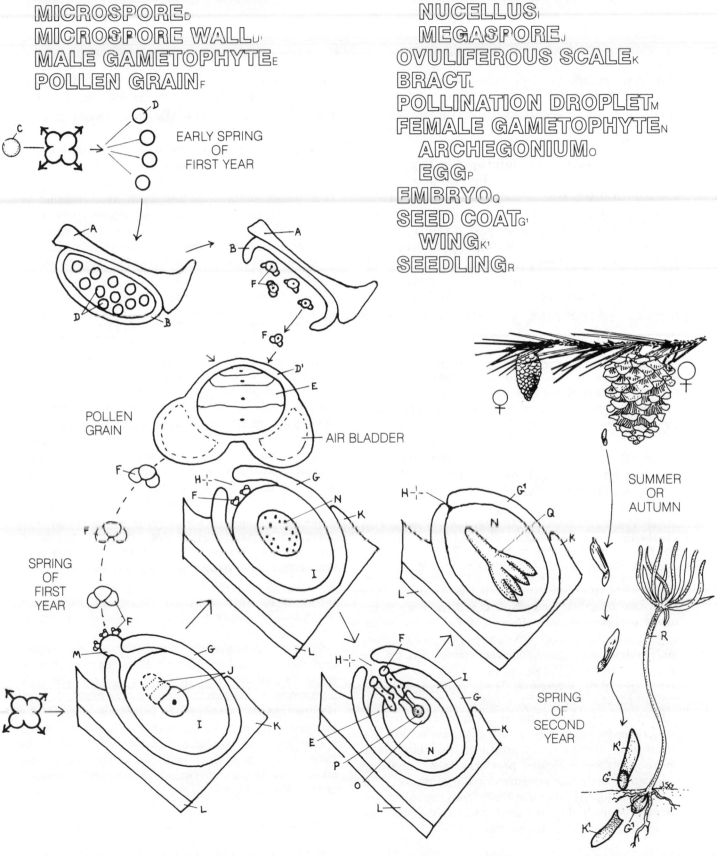

EARLY SPRING OF FIRST YEAR

POLLEN GRAIN

AIR BLADDER

SPRING OF FIRST YEAR

POLLEN GRAIN

SUMMER OR AUTUMN

SPRING OF SECOND YEAR

81

SEED PLANT REPRODUCTION

This plate compares some major reproductive features for the two types of seed plants, the gymnosperms (meaning "naked seeds") and the angiosperms (meaning "covered seeds").

Color the ovules (A), micropyles (B), pollen grains (H), and pistil (L) on the two diagrams at the top of the plate.

The exposed *ovules* of gymnosperms (a pine conifer, Coniferophyta, is illustrated) permit the transfer of *pollen grains* directly to the *micropyle*, where the pollination droplet (not shown), secreted through the *micropyle*, catches them. In flowering plants, Anthophyta, the *ovules* are completely surrounded and enclosed by a specialized sporophytic structure called a *pistil*. One or more *ovules* are located within each chamber, or *locule*, formed by the *pistil*. Angiosperm *pollen grains* are deposited on a specialized, often sticky, surface at the top of the *pistil* and not in the immediate vicinity of the *micropyle* because the *micropyle* is not exposed to permit direct transfer, as in gymnosperms.

Color the integument (C), nucellus (D), and megagametophyte (E) on the two ovule diagrams across the middle of the plate.

The gymnosperm *ovule* has a single-layered *integument*, several cells thick, of sporophytic tissue that completely surrounds, except for the *micropyle*, the megasporangium or *nucellus*. In gymnosperms, the *nucellus* produces from one (in pines) to a few diploid megasporocytes (not shown) that, upon meiosis, each form one functional megaspore and three nonfunctional megaspores (not shown). A functional megaspore undergoes mitotic division to produce a mature multicelluar *megagametophyte* surrounded by the *nucellus*. Between the *nucellus* and the *integument*, in the area of the *micropyle*, is a large chamber called the micropylar chamber (not colored).

In most angiosperms, the *ovules* have a two-layered *integument* and a very small chamber behind the *micropyle*. The *nucellus* produces only one megasporocyte that typically produces one functional megaspore and three nonfunctional megaspores. The *megagametophyte* that develops from the functional megaspore is highly reduced and consists of only seven cells in most angiosperms.

On the third row of diagrams, color the previously listed features and the archegonia (F), eggs (G), microgametophyte (I), tube nucleus (J), and sperm (K).

Though the gymnosperm *megagametophyte* is a single multinucleate cell in its early stages of development (not shown), at maturity, it consists of a multitude of uninucleate haploid cells. Each *megagametophyte* produces two or more *archegonia* completely imbedded within the *megagametophyte* with their neck region oriented toward the micropylar chamber (not shown). Gymnosperms produce typical *archegonia* with a swollen basal chamber that contains a single *egg* and a narrow neck region.

The seven-celled angiosperm *megagametophyte* has six uninucleate haploid cells and a large central cell containing two haploid nuclei called polar nuclei. The single *egg* is flanked by two cells that are thought to be a highly reduced *archegonium*.

In gymnosperms, dehydration of the pollination droplet pulls attached *pollen grains* through the *micropyle*, into the micropylar chamber, and in direct contact with the *nucellus*. Prior to germination, gymnosperm *microgametophytes* within the *pollen grain* consist of four to several cells (not shown). Upon germination, the *microgametophyte* forms a pollen tube that grows through the *nucellus* toward the *megagametophyte*. Within the pollen tube of the gymnosperm *microgametophyte* just prior to fertilization are several vegetative haploid nuclei (not separately colored), a *tube nucleus* at the apex, and two *sperm* located just behind the *tube nucleus*. Most gymnosperms produce nonmotile *sperm*, but a few, such as cycads and ginkgo, produce *sperm* motile by numerous cilia.

Prior to germination, most angiosperm *microgametophytes* are two celled. Upon germination on the surface of the *pistil*, the *microgametophyte* sends a pollen tube through the tissue of the *pistil*, through the *micropyle*, through the *nucellus*, and to the *megagametophyte*. At maturity, the pollen tube of the angiosperm *microgametophyte* contains only the *tube nucleus* at the growing end of the pollen tube and two nonmotile *sperm*. All angiosperm *sperm* are nonmotile.

OVULE AND MICROGAMETOPHYTE STRUCTURE.

GYMNOSPERMS ★

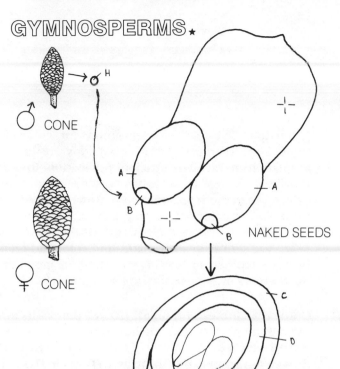

CONE ♂

CONE ♀

NAKED SEEDS

OVULE

ANGIOSPERMS ★

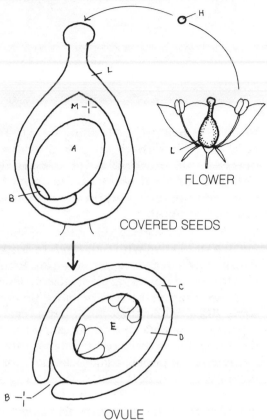

FLOWER

COVERED SEEDS

OVULE

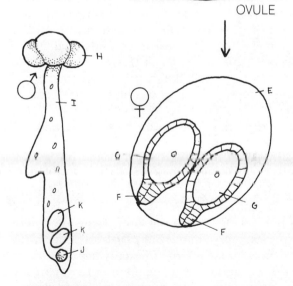

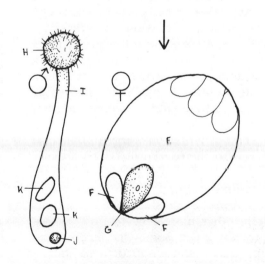

OVULE_A

OVULEA
MICROPYLEB
INTEGUMENTC
NUCELLUSD
MEGAGAMETOPHYTEE
ARCHEGONIUMF
EGGG

POLLEN GRAINH
MICROGAMETOPHYTEI
TUBE NUCLEUSJ
SPERMK
PISTILL
LOCULEM-¦-

82

BASIC FLOWER STRUCTURE

Color the pedicel (A)/peduncle (A¹) and receptacle (B) on the three large longitudinal section diagrams of flowers. Color the sepals (C) and petals (D) of the flowers on the upper half of the plate and the tepals (E) on the flower at the bottom of the plate.

The flower is a specialized branch, or shoot, consisting of a short stem with four kinds of parts. The individual flower parts are derived from highly specialized leaves. In flowering plants that produce an inflorescence (clusters of two or more flowers, such as some roses, geraniums, lilies, etc.), each flower is supported by a stalklike stem called the *pedicel.* In flowering plants that produce solitary flowers (such as some roses, poppies, and magnolia, for example), the stalklike stem is called a *peduncle.* In addition, the stem supporting an inflorescence is called a *peduncle* (see Plate 91). In any case, the flower-bearing stem is terminated by an enlarged end, the *receptacle,* to which the individual flower parts are attached. The tight clustering of flower parts is due to little or no elongation of the *receptacle.* In most flowers, all parts of one kind are arranged in a distinct group, called a floral series. For example, the *calyx,* one floral series, is made up of one kind of flower part, the *sepals.* Since each flower has four kinds of parts, each has four distinct floral series.

The lower two floral series, the *calyx* and the *corolla,* are referred to as the two sterile series because they do not function directly in sexual reproduction. The first, or lowermost, floral series, is the *calyx.* While the flower is in the bud stage, the *calyx* functions to protect the other, more delicate, flower parts. The *sepals* are usually green, somewhat leaflike, and relatively tough, such as the *sepals* of roses, but in some flowers, such as most lilies, the *sepals* are delicate and colorful at maturity and closely resemble the *petals* of the second floral series.

The second floral series, called the *corolla,* consists of individual flower parts called *petals.* A major function of the *corolla* in most flowers is attracting specific animals, mostly insects, that will efficiently transfer pollen from flower to flower, a process called pollination. The shape, background color, color patterns, and other features of the *petals* are usually adaptations to facilitate efficient pollination. In most flowers, the *petals* are usually larger and more colorful than the *sepals* when the flower is open.

The first two floral series, the *calyx* and the *corolla,* are collectively called the perianth. If the *sepals* and *petals* are similar in size, shape, and color and grade into one another, such as in magnolias, the individual flower parts of the perianth are called *tepals.*

Color the stamens (F) on all flower diagrams and the filament (G) and anther (H) of the single boxed stamen. Also color the pistils (I) on all flower diagrams and the stigma (J), style (K), and ovary (L) of the single boxed pistil.

The upper two floral series, the androecium and the gynoecium, are referred to as the two fertile series because they function directly in sexual reproduction. The third floral series, called the androecium, which means "male home," consists of individual flower parts called *stamens.* In most flowers, each *stamen* consists of a thin, stalklike *filament* and an enlarged end, the *anther,* consisting of one to four chambers within which pollen is produced. Pollen, which is released by rupture of the mature *anther,* contains the male gametes.

The fourth floral series, called the gynoecium, which means "female home," consists of individual floral parts called *pistils. Pistils* may be derived from a single specialized leaf or two or more specialized leaves that are fused into one unit. Depending upon plant species, one to numerous *pistils* are found within the gynoecium. Each *pistil* usually has three parts: *stigma, style,* and *ovary.* The apex of the *pistil,* the *stigma,* is often branched or lobed and has a receptive surface that exudes a sticky substance to which pollen adheres. The narrowed, necklike portion of the *pistil* below the *stigma* is the *style,* but in some flowers, it is reduced or lacking. The enlarged base of the *pistil,* the *ovary,* contains one or more chambers in which one to many thousands of female gametes are produced.

BASIC FLOWER STRUCTURE.

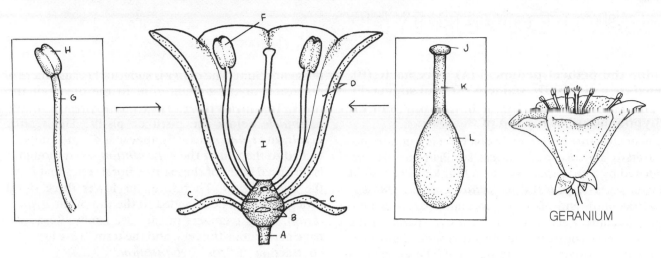

GERANIUM

FLOWER PARTS.
PEDICELA/PEDUNCLEA¹
RECEPTACLEB
PERIANTH ●
  CALYXC( )
    SEPALC
  COROLLAD( )
    PETALD
    TEPALE
ANDROECIUM ●
  STAMENF
  FILAMENTG
  ANTHERH
GYNOECIUM ●
  PISTILI
    STIGMAJ
    STYLEK
    OVARYL

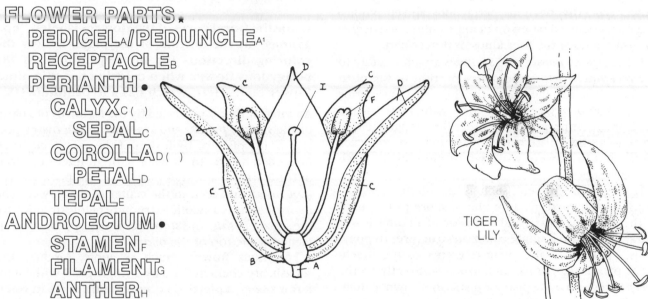

TIGER
LILY

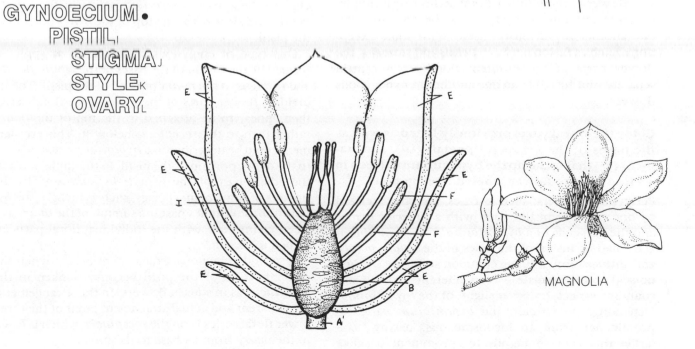

MAGNOLIA

83

BASIC FLOWER TYPES

Color the pedicel/peduncle (A), receptacle (B), sepals (C), petals (D), stamens (E), and stigma (F), style (G), and ovary (H) of the flower labeled "hypogynous" at the top of the plate.

Flower structure is extremely diverse. Flowers are supported by a *pedicel* or *peduncle* (see Plates 82, 91) terminated by a *receptacle* to which four kinds (series) of flower parts—*sepals* (calyx), *petals* (corolla), *stamens* (androecium), and *pistils* (gynoecium)—are attached. The attachment sequence of the flower parts on the *receptacle* is constant, but fusion among the flower parts may obscure this arrangement. Three general flower types (hypogynous, perigynous, and epigynous) can be recognized based on *ovary* position and degree of fusion among the four kinds of flower parts.

If the *sepals, petals,* and *stamens* attach directly to the *receptacle* below the *ovary* portion of the pistil with no fusion to the *ovary,* the flower has a superior *ovary.* The *ovary* is seen as the swollen base of the pistil above (superior to) the other flower parts. Flowers with a single pistil may or may not have a superior *ovary,* but in flowers with two or more pistils, each *ovary* is superior.

Hypogynous flowers, such as geraniums or carnations, are characterized by one or more pistils, each with a superior *ovary,* and no fusion of all three lower floral series into a single common structure. Hypogynous, "below female," is in reference to the attachment of *sepals, petals,* and *stamens* directly to the *receptacle* below the pistil or pistils in the gynoecium. The flower parts within a floral series (for example, the *petals* within the corolla) may be separate from one another or variously fused. With some exceptions, such as fusion between *petals* and *stamens,* the flower parts of the different floral series remain separate (not fused) from one another in hypogynous flowers.

Color the floral parts previously listed as well as the bases of the sepals (C¹), petals (D¹), and stamens (E¹) that make up the hypanthium (I) (seen in section) on the flower labeled "perigynous."

Perigynous flowers, such as roses, are characterized by one or more pistils, each with a superior *ovary,* and the presence of a *hypanthium.* The *hypanthium* is formed by fusion of the bases of the *sepals, petals,* and *stamens* into a single common structure. Perigynous ("around female") is in reference to the surrounding aspect, however slight, of the *hypanthium.* Depending on species, the *hypanthium* varies in prominence from an inconspicuous, narrow, collarlike rim, as in cinquefoils, to a prominent, flasklike or vaselike cup, as in roses, that completely surrounds the gynoecium. The *ovary* is superior because there is no fusion of the *hypanthium* to the *ovary* of the pistil. The *hypanthium* is attached directly to the *receptacle* below the pistil or pistils. The *sepals, petals,* and *stamens* usually appear as distinct units attached to the rim of the *hypanthium,* even though it is the fused bases of these three floral series that form the *hypanthium.* Therefore, the lower three floral series appear to be attached to the rim of the *hypanthium.* Many members of the rose family, Rosaceae, have perigynous flowers, and the term "rose hip" is a contraction of "rose *hypanthium.*"

Color the floral parts of the flowers labeled "epigynous" at the bottom of the plate. Follow the coloring directions for the hypanthium of the perigynous flower when coloring the epigynous flower on the left, which has a hypanthium (I).

An inferior *ovary* is formed by the fusion of the bases of the *sepals, petals,* and *stamens* to the *ovary* portion of the pistil or by fusion of the *receptacle* to an imbedded *ovary.* In flowers with an inferior *ovary,* only the *stigma* and *style* portions of the pistil are free (not fused) and seen in the center of the flower. The *ovary* is seen as a swollen structure below (inferior to) the *sepals, petals,* and *stamens,* which appear to be attached at the top of the *ovary.*

Epigynous flowers, such as apple, fuschia, and squash, are characterized by a single pistil with an inferior *ovary.* Epigynous ("upon female") is in reference to the apparent insertion points of the *sepals, petals,* and *stamens* at the top of the *ovary* portion of the pistil.

One type of epigynous flower, such as apple, is formed by the fusion of the bases of the *sepals, petals,* and *stamens* to the *ovary* portion of the pistil. The individual flower parts of the lower three floral series then appear to be attached to the top of the *ovary* rather than to the *receptacle* below it. This arrangement could be described as a *hypanthium* that is fused to the *ovary* portion of the pistil. In the apple, most of the edible part of the fruit is *hypanthium.* In the flowers of some plant species, such as fuchsia, the *hypanthium* continues past the summit of the *ovary* as a free *hypanthium* with the individual floral parts attached to its rim.

A second epigynous arrangement occurs when the *ovary* portion of the pistil becomes sunken in the *receptacle,* as in squash flowers. In this arrangement, the apparent and actual attachment point of the three lower floral series is on the *receptacle,* which is fused to the *ovary* from its base to its summit.

FLOWER TYPES.

FLOWER PARTS★
PEDICEL/PEDUNCLE A
RECEPTACLE B
SEPAL C
PETAL D
STAMEN E
PISTIL ★
   STIGMA F
   STYLE G
   OVARY H
HYPANTHIUM I
   BASE OF:
   SEPAL C¹
   PETAL D¹
   STAMEN E¹

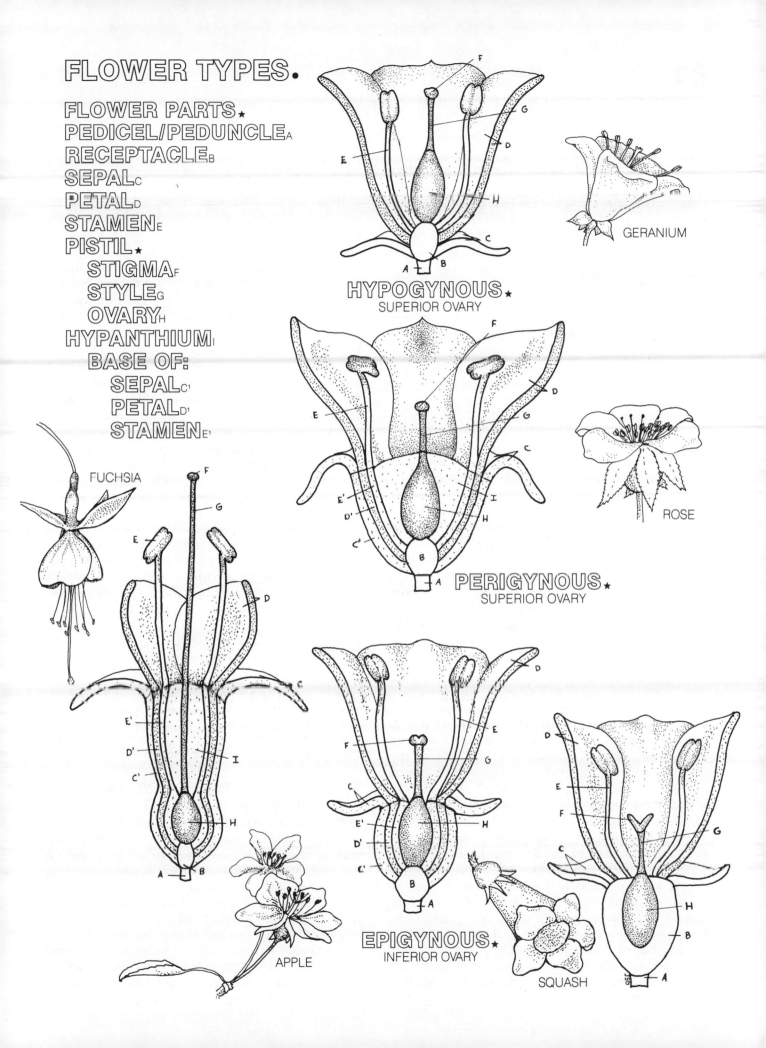

HYPOGYNOUS ★
SUPERIOR OVARY

GERANIUM

PERIGYNOUS ★
SUPERIOR OVARY

ROSE

FUCHSIA

APPLE

EPIGYNOUS ★
INFERIOR OVARY

SQUASH

Starting with the diagram depicting a leaflike carpel at the top center of the plate, color the carpels (A), placental ridges (B), ovules (C), and stigma (D), style (E), and ovary (F) of the simple pistils (H) on the upper right quarter of the plate.

The gynoecium consists of all the pistils, one or more, depending on plant species, present in a single flower. A pistil is formed by a single *carpel* or by two or more fused *carpels*. A *carpel,* the megasporophyll or megaspore-forming leaf of the gynoecium, is a highly specialized leaf that has two strips of fertile tissue, called *placental ridges,* that produce *ovules.* The two *placental ridges* lie along the edges of the *carpel.* Each *placental ridge* may produce numerous *ovules,* and each *ovule* may mature into a seed.

A pistil derived from a single *carpel* is called a *simple pistil.* The *stigma* and *style* of the pistil are formed by the tip and narrow upper portions of the *carpel;* the *ovary* of the pistil is formed by the expanded, *ovule*-bearing portion of the *carpel.* Inrolling and fusion of the *carpel* margins form one *locule,* or ovary chamber, in a *simple pistil.* A single *ovule*-bearing *placental ridge,* running the length of the *locule,* is formed by fusion of the two marginal *placental ridges.* This type of placental arrangement is called marginal placentation. A pea pod is a pistil formed from a single *carpel.* The *stigma, style,* and *ovary* portions of the pistil are labeled in the center illustration of a whole pea pod. The longitudinal section (l.s.) of the pea pod and the laid-open pea pod reveal the *placental ridge,* which is split into two strips when the pea pod is opened, and the peas, which are the maturing *ovules.*

Color the labeled structures, previously listed, on the diagrams of marginal carpel fusion in compound pistils (I) along the left side of the plate. Stop when you finish with the diagrams depicting "parietal placentation." The arrows indicate a proposed evolutionary sequence.

Compound pistils are derived from the fusion of two or more *carpels* into a single pistil. The manner and degree of fusion between adjacent *carpels* in compound pistils varies to produce a number of different *carpel* arrangements. In one arrangement, here labeled "marginal fusion," fusion occurs between the margins of adjacent *carpels* only. This produces an *ovary* with a single *locule.* In parietal placentation, the number of *placental ridges,* or rows of *ovules,* corresponds to the number of fused *carpels* since the adjacent *placental ridges* from adjoining *carpels* fuse to form a single *placental ridge.* This type of pla-

cental ridge arrangement, or placentation, is called parietal placentation, in reference to the peripheral position of the *placental ridges.* Members of the squash family, such as cantaloupes, squashes, and gourds, have parietal placentation.

Color the labeled structures, previously listed, on the diagrams of septal carpel fusion in compound pistils (I) in the lower right quadrant of the plate. Stop when you finish with the diagrams depicting "free central placentation."

In a second arrangement of fused *carpels,* here labeled "septal fusion," the margins of the individual *carpels* are inrolled and fused, as in a *simple pistil.* Each *carpel* forms an individual *locule* and a single *placental ridge,* but fusion between adjacent *carpels* is extensive. Each *locule* is separated from the adjacent *locules* by a partition, called a septum (not separately colored), formed by fusion of adjacent *carpel* walls. In axile placentation, the single *placental ridge* formed by each *carpel* runs the length of the central axis of the pistil. The arrangement of *placental ridges* in this manner is called axile placentation, in reference to the placement of the *placental ridges* along the central axis. In some pistils, such as in apples, partitions, called septa, divide the *ovary* into two or more *locules* corresponding to *carpel* number. Apples have five septa, which, along with the ovary wall, form a hard core, and five ovary chambers, or *locules,* which contain the *ovules* (seeds). In other plants, the septa are absent so that the *ovary* has a single *locule.* If the central *ovule*-bearing column is not attached to the top of the *ovary,* a free-standing column of *ovule*-bearing tissue, composed of *placental ridges* and associated *carpel* tissue, is produced. This arrangement is called free central placentation and is found in all members of the carnation family.

Color the diagrams illustrating "basal placentation" at the bottom of the plate.

Both marginal and septal fusion may be the evolutionary basis for basal placentation in which the *ovule*-bearing tissue has been reduced to a small mound at the base of the *ovary* chamber, a single *locule* is typically present, and *ovules* are reduced to one or a few. One-seeded fruits, such as peaches, cherries, plums and mangoes, have basal placentation. The thin arrows between the diagrams of *compound pistils* indicate proposed evolutionary sequences from many-seeded pistils to one-seeded pistils through reduction, or loss, of parts. Basal placentation may be derived from marginally or septally fused *carpels.*

PISTIL, CARPEL AND OVARY.

CARPEL (MEGASPOROPHYLL)A
PLACENTAL RIDGEB
OVULEC
PISTIL ★
STIGMAD
STYLEE
OVARYF
LOCULEG

LEAFLIKE CARPEL

MARGINAL FUSION ★

COMPOUND PISTIL

SIMPLE PISTILH

PEA POD

MARGINAL PLACENTATION

SEPTAL FUSION ★

AXILE PLACENTATION

PARIETAL PLACENTATION

FREE CENTRAL PLACENTATION

CENTRAL FREE-STANDING COLUMN

BASAL PLACENTATION

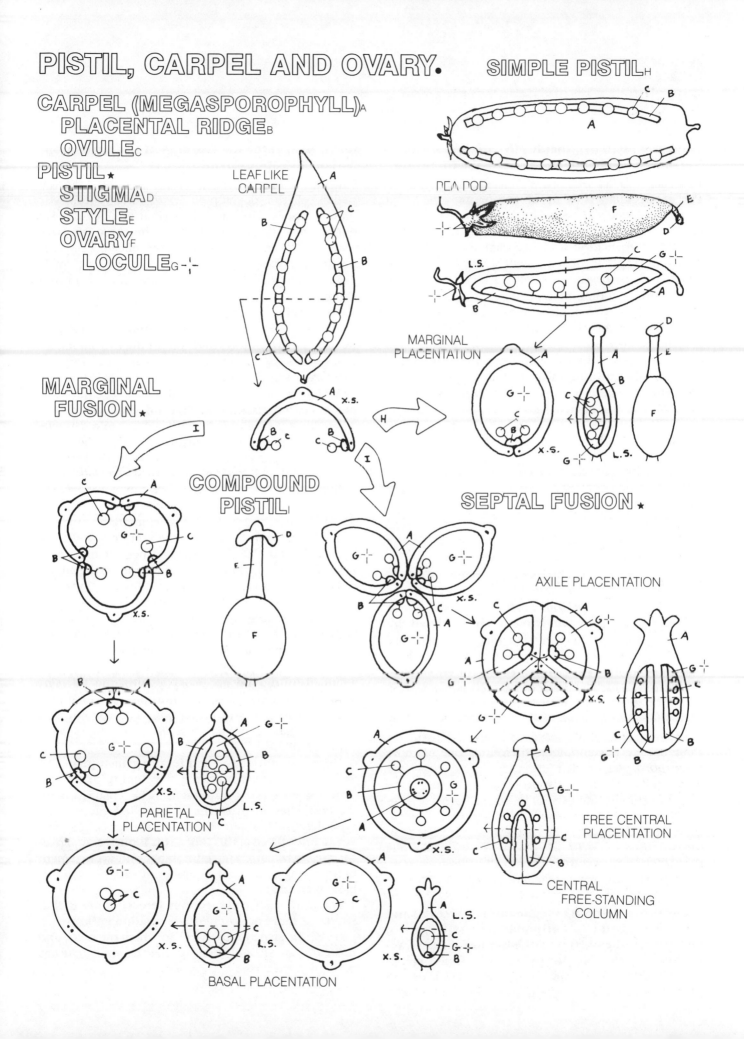

85

OVULE DEVELOPMENT

Color the pistil (A), locules (B), placental ridges (C), funiculi (D), and ovules (E) on the three large diagrams across the top of the plate.

The gynoecium, made up of one or more *pistils* (a flower with one *pistil* is illustrated), is the floral series in which megasporogenesis, or megaspore (a female meiospore) formation, and megagametogenesis, or female gametophyte development, occur. Each *pistil* consists of one or more carpels (three are illustrated) and a corresponding number of *locules* or, in some compound *pistils*, a single, common *locule*. *Placental ridges* bearing the *ovules* supported by a short stalk, the *funiculus,* are located within the *locules.* Though illustrated as a distinct structure, a *placental ridge* is anatomically indistinguishable from carpel tissue of the *pistil.* Because each *placental ridge* is actually formed by fusion of two *placental ridges,* the *ovules* on a *placental ridge* are frequently arranged in two rows.

Color the ovule primordium (F), integuments (G), nucellus (H), micropyle (I), megasporocyte (J), functional megaspore (K), abortive megaspores (L), meiosis symbol, and previously listed structures on the diagrams in the center of the plate illustrating megaspore formation.

Each *ovule* begins development as a mound of tissue, an *ovule primordium,* on a *placental ridge.* As the *ovule* develops, its various anatomical areas become differentiated. Two *integument* layers (fused together and colored as one structure) grow upward from the *funiculus* end of the *ovule* to form a flask-shaped protective cover fused to the nucellar tissue, or *nucellus,* which it almost completely surrounds except for a small pore, the *micropyle,* at the distal end of the developing *ovule.*

The *nucellus,* which serves as a nutritive tissue for gametophyte development, surrounds one or more *megasporocytes,* which are the diploid cells that undergo meiosis in megasporogenesis (the formation of megaspores). Meiosis of a *megasporocyte* produces one viable megaspore, the *functional megaspore,* and three *abortive megaspores,* smaller than the *functional megaspore,* that are reabsorbed while the *functional megaspore,* typically the megaspore farthest from the *micropyle,* enlarges.

Color the developing megagametophyte (M) and the central cell (N), antipodal cells (O), egg (P), and synergid cells (Q) of the mature megagametophyte as well as the previously listed structures on the diagrams along the bottom of the plate illustrating female gametophyte develop-

ment. Also color the features of the anatropous and orthotropous ovules.

Except for the haploid *functional megaspore,* all tissues of the developing *ovule* are sporophytic diploid tissues. The uneven distribution of cytoplasm during meiosis, reabsorption of the three *abortive megaspores,* and absorption of the *nucellus* provide nutritional reserves for megagametogenesis (female gametophyte development) and seed formation (not shown).

To complete megagametogenesis, a series of three synchronous mitotic divisions, without cytokenesis, occur. The first mitotic division produces a binucleate *megagametophyte.* In the second mitotic division, these two haploid nuclei divide to produce a *megagametophyte* with four haploid nuclei. The third and last synchronous mitotic division produces a one-celled *megagametophyte* with eight haploid nuclei that migrate to specific positions within the eight-nucleate *megagametophyte.* Cytokenesis forms a multicellular *megagametophyte* of seven cells.

In migration and cytokenesis, two nuclei migrate to the center of the *megagametophyte* to form a *central cell* that is binucleate. Three nuclei migrate to the end opposite the *micropyle,* the chalazal end, and form three cells of undefined purpose called the *antipodal cells.* The remaining three nuclei migrate to the micropylar end and form an *egg* and two flanking *synergid cells,* a complex called the "egg apparatus." At maturity, the female gametophyte or *megagametophyte* consists of a binucleate *central cell,* three *antipodal cells,* two *synergid cells,* and a single female gamete, the *egg.* Each haploid female gametophyte is surrounded by diploid accessory tissues, the *nucellus* and the *integuments,* of the sporophyte, which provide nourishment and protection. The combination of female gametophyte, *nucellus,* and *integuments* forms an *ovule.*

Megaspore formation and female gametophyte development occur within the developing *pistil* of a flower while it is in the bud stage. The female gametophyte reaches maturity, the seven-celled stage for the type described, about the time a flower opens. The *egg* is unfertilized at this time. This is one general plan for *megagametophyte* cell number and arrangement in flowering plants, but a number of variations of this basic plan exist.

For clarity, an orthotropous (erect) *ovule* is illustrated. A more prevalent *ovule* position in flowering plants is the anatropous *ovule,* in which the *ovule* and *funiculus* are strongly bent so that the *micropyle* lies near the *funiculus.*

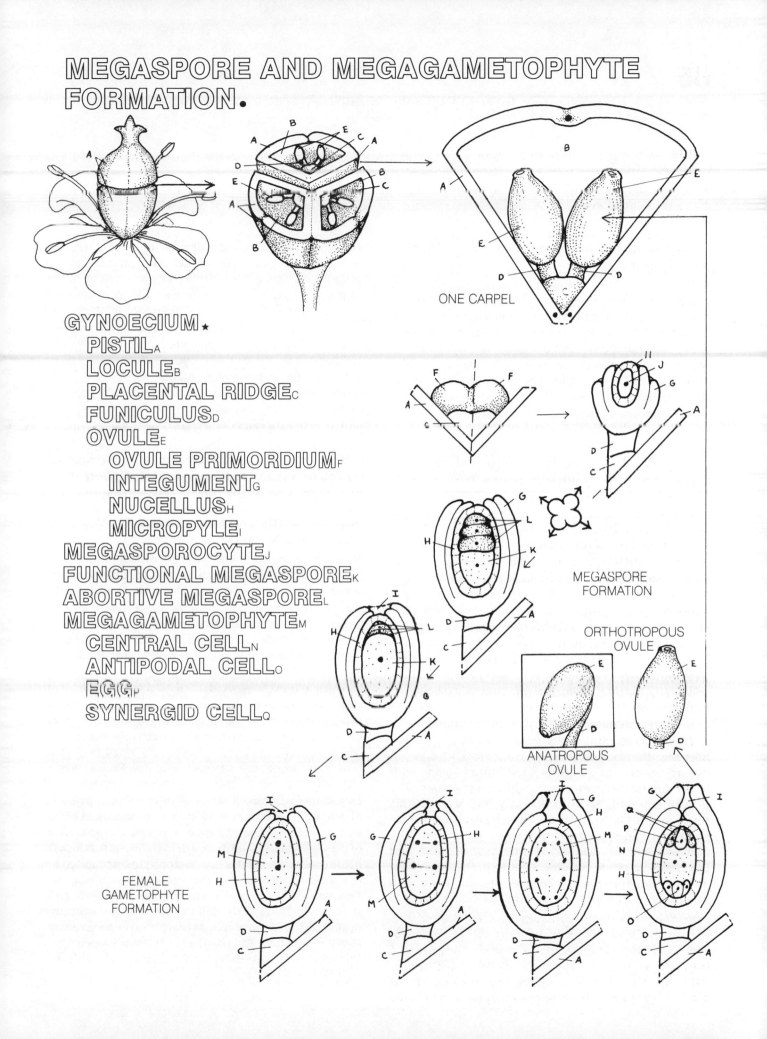

MEGASPORE AND MEGAGAMETOPHYTE FORMATION.

ONE CARPEL

GYNOECIUM ★
   PISTIL A
   LOCULE B
   PLACENTAL RIDGE C
   FUNICULUS D
   OVULE E
     OVULE PRIMORDIUM F
     INTEGUMENT G
     NUCELLUS H
     MICROPYLE I
MEGASPOROCYTE J
FUNCTIONAL MEGASPORE K
ABORTIVE MEGASPORE L
MEGAGAMETOPHYTE M
   CENTRAL CELL N
   ANTIPODAL CELL O
   EGG P
   SYNERGID CELL Q

MEGASPORE FORMATION

ORTHOTROPOUS OVULE

ANATROPOUS OVULE

FEMALE GAMETOPHYTE FORMATION

86

Color the flower development sequence across the top of the plate.

Flowering plants, being heterosporous, produce meiospores that are either *microspores* (male) or *megaspores* (female). Consequently, the gametophyte that develops from a single meiospore produces either male gametes or female gametes. Meiosis usually occurs very early in flower development, while the flower is still a small bud. The bud continues to enlarge as it develops. At anthesis, flower opening, the *sepals* and *petals* open to expose the *stamens* or *pistil* or *pistils*.

Color the diagrams illustrating microsporogenesis and microgametogenesis across the middle of the plate.

Microsporogenesis is the process by which a diploid cell, the *microsporocyte,* undergoes meiotic division to produce haploid *microspores. Microspores* develop into the male gametophyte. Microsporogenesis occurs in the stamens, the "male portion" of a flower. Each stamen consists of a *filament* and *anther,* which has one to four chambers, called pollen sacs or anther sacs. Each anther sac contains numerous *microsporocytes,* also called pollen mother cells. Through meiosis, each *microsporocyte* produces four functional *microspores.* The *microspores* produce a thick cell wall, the *microspore wall,* that is resistant to desiccation.

Microgametogenesis, which is microgametophyte (male gametophyte) development, begins with the first mitotic division of a *microspore.* This division occurs within the confines of the *microspore wall* while the *microspore* is still within the *anther.* The flower is usually still in the early bud stage at this time. Additional wall layers are added to the *microspore wall* as it becomes the pollen grain wall. The combination of elaborated *microspore wall* and microgametophyte is called a pollen grain. In the first mitotic division, the microspore nucleus divides to form a two-celled microgametophyte consisting of a *vegetative cell* and a *generative cell.* The flower usually opens about this time. In most flowering plants, the pollen grains are released from the anther

sacs by rupture of the sac wall while they are in the two-celled stage. Maturation of the microgametophyte is completed outside the *anther.* Mitotic division of the *generative cell* to form two *sperm cells* completes the maturation of the microgametophyte. Thus, the fully developed microgametophyte consists of three haploid cells, one *vegetative cell,* and two *sperm cells,* surrounded by the modified *microspore wall.* Pollen grains in most flowering plants are carried from flower to flower in the two-celled stage of development, but some species (about 20 percent) release pollen at the three-celled stage.

Color the diagrams illustrating megasporogenesis and megagametogenesis across the bottom of the plate. Color the pistil (F), megasporocyte (M), nucellus (N), integument (O), funiculus (P), megaspores (R), and megagametophyte (S). Note that the locule (Q) is not colored.

Megasporogenesis, *megaspore* formation, and megagametogenesis, female gametophyte development, occur within the gynoecium, the "female home" of a flower. In review, prior to megasporogenesis, the ovary part of the *pistil* contains one to numerous developing ovules consisting of one or more diploid *megasporocytes,* or egg mother cells, surrounded by the *nucellus* and the two *integuments* and supported by the stalklike *funiculus.*

In megasporgenesis, meiosis of a *megasporocyte* produces one large functional *megaspore* and three smaller abortive *megaspores* that are reabsorbed. This usually occurs while the flower is a small bud. The functional *megaspore* becomes a developing *megagametophyte* with the first mitotic division of megagametogenesis. A series of three mitotic divisions in which cytokinesis does not occur produces an eight-nucleate *megagametophyte* or embryo sac. Migration of the eight nuclei coupled with cytokinesis produces a mature, seven-celled megagametophyte surrounded by the supportive tissues, the *nucellus,* and the *integuments* of each ovule. The *megagametophyte* consists of six uninucleate cells and one large, centrally positioned binucleate cell. At this time, the *megagametophyte* is mature and the flower is usually open.

GAMETOPHYTE DEVELOPMENT.

FLOWER★

PEDICEL/PEDUNCLE_A
SEPAL_B
PETAL_C

STAMEN★
FILAMENT_D
ANTHER_E
PISTIL_F

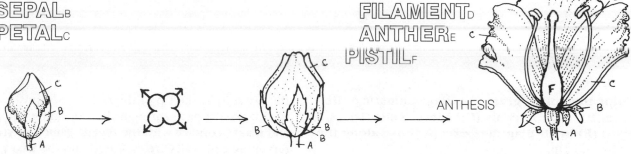

ANTHESIS

MICROSPOROGENESIS★
MICROSPOROCYTE_G
MICROSPORE_H
MICROSPORE WALL_I

MICROGAMETOGENESIS★
MICROGAMETOPHYTE★
VEGETATIVE CELL_J
GENERATIVE CELL_K
SPERM CELL_L

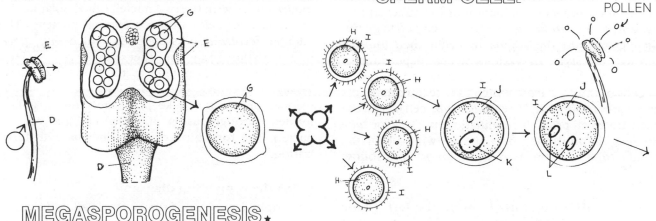

POLLEN

MEGASPOROGENESIS★
OVULE★
MEGASPOROCYTE_M
NUCELLUS_N
INTEGUMENTS_O

FUNICULUS_P
LOCULE_Q
MEGASPORE_R
MEGAGAMETOGENESIS★
MEGAGAMETOPHYTE_S

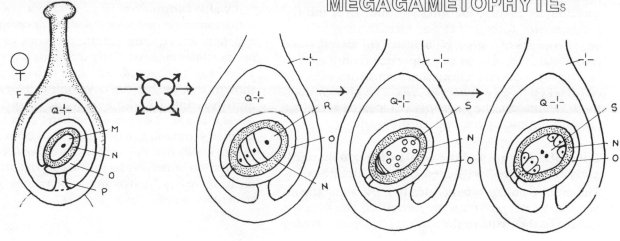

87
FLOWERING PLANT REPRODUCTION

Color the pollen grain (A), diagrammatic pollen tube (B), tube nucleus (C), sperm cells (D), and pistil (E) in all diagrams except those along the right margin.

A flower opens when all flower parts are at or near maturity. In many flowers, the androecium and gynoecium mature at different times to discourage the transfer of *pollen grains* from the anthers to the *pistil* or *pistils* of the same flower.

When mature, changes of various types (depending on species) occur in the stigma to make it receptive to *pollen grains.* Substances secreted by the stigma, such as enzymes, induce germination of the attached *pollen grains. Pollen tubes,* one from each *pollen grain,* then grow into the stigma to begin their journey through the style. By various means (depending on species), the *pollen tubes* grow among and appressed to cells of the style from which they receive nutrients and water for growth. The *pollen tube* cell (called the vegetative cell prior to germination) nucleus, or *tube nucleus,* is positioned near the growing tip of the *pollen tube,* and the two small *sperm cells* follow close behind.

Color the first pistil diagram on the left. Color the following two pistil diagrams, noting that many structures are not colored.

The locule, or locules, of the *pistil* contain one to numerous ovules, each supported by a *funiculus.* Each fully mature ovule consists of a megagametophyte (female gametophyte or embryo sac) that is surrounded by two nutritive and protective diploid sporophytic tissues, the *nucellus* (nutritive) and the *integuments* (protective). A mature megagametophyte consists of a large, binucleate *central cell,* three *antipodal cells,* and an egg apparatus (considered to be a highly reduced archegonium) consisting of an *egg cell* flanked by two *synergid cells.* The pore through the *integuments,* the *micropyle,* is located at the *egg cell* end, or micropylar end, of the ovule.

Color the features of the ovule diagram immediately below the "withered flower" diagram.

The minute *pollen tubes* typically grow along the inner wall of the *pistil,* up the *funiculus,* and over the surface of the ovule to the *micropyle.* Upon reaching

the *micropyle,* the tip of the *pollen tube* pushes into it, dissolves a path through the remaining *nucellus,* and makes contact with the female gametophyte. Intervening cell walls are ruptured and the two *sperm cells,* along with the *tube nucleus,* are released into one *synergid cell.* The receiving *synergid cell* transfers one *sperm* to the *egg* and ruptures to release the second *sperm* into the *central cell.* The ruptured *synergid cell,* along with the other *synergid cell* and the *antipodal cells,* soon degenerates. The *sperm cell nucleus* transferred to the *egg cell* fuses with the *egg* nucleus to form the *zygote.* The other *sperm cell nucleus* fuses with the two nuclei, called polar nuclei, of the *central cell* to form a triploid nucleus. This "double fertilization" system, in which one *sperm* fuses with the *egg* and the other with the two nuclei of the *central cell,* is unique to the flowering plants. Though it is called "double fertilization " because of the two unions that take place, only one *zygote* is formed (through fusion of *sperm cell* and *egg cell* nuclei). The formation of a seed begins with fertilization.

Color the remaining diagrams.

The triploid nucleus of the *central cell* undergoes a series of mitotic divisions to produce the *endosperm,* a tissue used as a nutrient source by the developing *embryo.* If no cell walls form, the *endosperm* is liquid, but it becomes solid if cell wall formation (not shown) occurs. In the coconut, the solid, white, meaty portion is cellular *endosperm;* the liquid portion is free nuclear (having many nuclei and without cell walls) *endosperm.*

Concurrently with *endosperm* development, the *zygote* is undergoing mitotic divisions to produce a multicellular *embryo.* Early divisions form a compact *embryo* and a chain of cells, called the suspensor (not shown), that pushes the developing *embryo* into the nutritive triploid *endosperm* tissue. Maturation of the ovule into a seed is complete when a fully formed *embryo* lies dormant within the *integuments* of the ovule, which become the hardened *seed coat.* During maturation of the ovules into seeds, the *pistil* dramatically increases in size and, often along with other structures, forms a fruit.

POLLINATION AND FERTILIZATION.

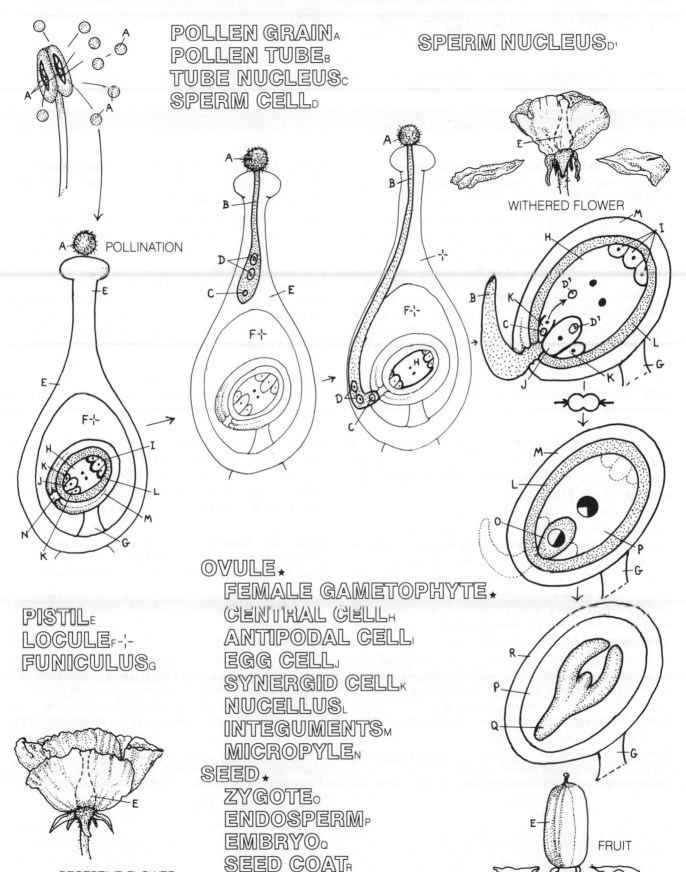

POLLEN GRAINA
POLLEN TUBEB
TUBE NUCLEUSC
SPERM CELLD

SPERM NUCLEUSD¹

POLLINATION

WITHERED FLOWER

PISTILE
LOCULEF -¦-
FUNICULUSG

OVULE★
  FEMALE GAMETOPHYTE★
  CENTRAL CELLH
  ANTIPODAL CELLI
  EGG CELLJ
  SYNERGID CELLK
  NUCELLUSL
  INTEGUMENTSM
  MICROPYLEN
SEED★
  ZYGOTEO
  ENDOSPERMP
  EMBRYOQ
  SEED COATR

RECEPTIVE FLOWER

FRUIT

Color the cotyledons (A) and root (B) and stem (C) regions of the two embryo types at the bottom of the plate.

Based on a number of morphological and anatomical characteristics, the Anthophyta, or Magnoliophyta, are separated into two classes: the Monocotyledoneae (monocots), or Liliatae, and the Dicotyledoneae (dicots), or Magnoliatae. The two classes derive their names from the number of *cotyledons* (embryonic seed leaves) present on the embryo. Monocots typically have one *cotyledon* and dicots typically have two. The remaining external morphological features of the embryos, the embryonic *root,* or radicle, and the embryonic *stem,* or epicotyl, differ little between the two classes.

Color the roots (B), xylem (D), cambium (E), and phloem (F) where present on the root systems and root cross sections below the soil line at the bottom of the plate.

Root system structure also provides a clue to class. Monocots usually have no main *root.* Instead, they have a fibrous *root* system consisting of numerous major *roots* that arise from the base of the plant. As seen in the cross section, monocot *roots* usually have distinct and separate bundles of *xylem* and *phloem* that alternate with one another in one or more peripheral rings. No vascular *cambium* is present to produce secondary *root* growth.

Dicots often have a tap *root* system composed of a single main *root* with many secondary *roots* branching from it. In most dicots, the vascular tissues form a central bundle in the center of the *root.* The *xylem* forms a central, often star-shaped, core that is surrounded by a vascular *cambium* that produces secondary growth. Bundles of *phloem* occupy the spaces between the lobes of *xylem.*

Color the xylem, cambium, and phloem on the stem (C) cross sections and the stems and leaves (G) on the two plants.

Monocot *stems* typically have numerous bundles of vascular tissue, consisting of both *xylem* and *phloem,* that are scattered throughout the *stem,* as seen in the *stem* cross section. In some monocots, the vascular bundles are arranged in concentric rings. No vascular *cambium* is present to produce secondary growth, and most monocots are nonwoody herbaceous plants (for example, lilies). The *leaves* of monocots are frequently bladelike, sessile, and sheathing at the base. The *leaf* venation pattern is typically parallel. Dicot *stems,* prior to secondary growth in those that are woody, have discreet bundles, or strands, of vascular tissue. The *xylem* is positioned on the inside and the *phloem* is on the outside of each bundle. In woody plants, those that produce secondary growth, a vascular *cambium* develops between the *xylem* and *phloem* and between the vascular bundles to form a vascular *cambium* cylinder in the *stem.* Both herbaceous and woody plant forms are well represented in the dicots.

Color the dicot sepals (H), monocot sepals (I), petals (J), stamens (K), and pistils (L) on the two flower types. Use green for the dicot sepals.

Monocot flowers are usually three-merous (that is, the flower parts of the various floral series are present in multiples of three). The typical monocot flower consists of three *monocot sepals,* which are frequently petallike, three *petals,* three or six *stamens,* and a tricarpellate compound *pistil.* Dicot flowers are usually either four-merous or five-merous, with flower parts in multiples of four (as in the evening primrose family) or multiples of five (as in the rose family). The five-merous flower illustrated has five *dicot sepals,* which are usually green and leaflike, five *petals,* five *stamens,* and a five-carpellate compound *pistil.*

CHARACTERISTIC FEATURES.

COTYLEDON A   LEAF G
ROOT B   DICOT SEPAL H
STEM C   MONOCOT SEPAL I
XYLEM D   PETAL J
CAMBIUM E   STAMEN K
PHLOEM F   PISTIL L

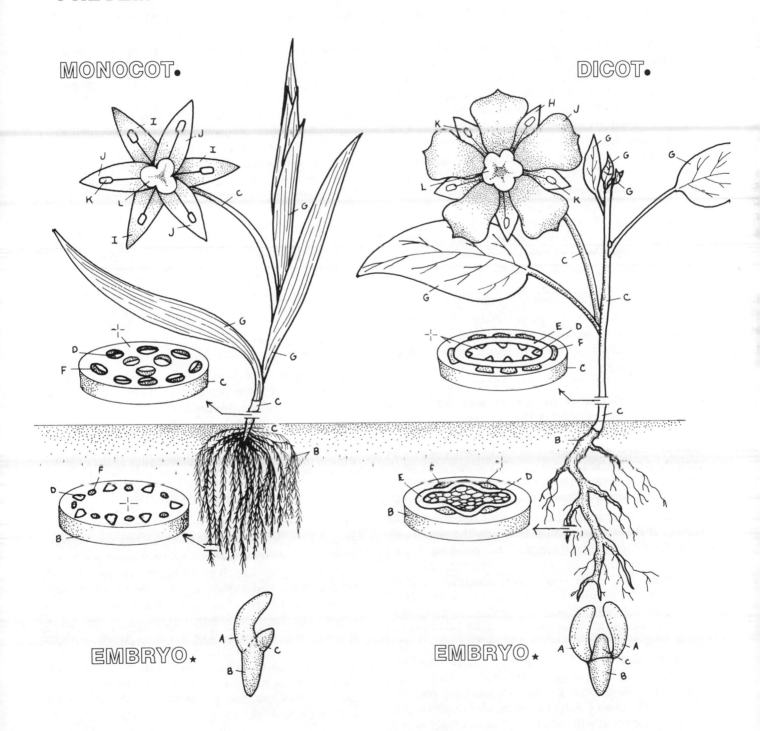

MONOCOT.

DICOT.

EMBRYO ★

EMBRYO ★

89

FLORAL SPECIALIZATION

Flower morphology ranges from relatively unspecialized to highly specialized. Most floral specialization is due to the evolution of adaptations that enhance reproductive efficiency. Three major aspects of flower specialization are fusion, reduction, and ornamentation of flower parts. Fusion may occur between two or more flower parts of the same floral series (connation), between the flower parts of different floral series (adnation), or a combination of connation and adnation. A lack of fusion is considered unspecialized, whereas fusion is considered specialized. Reduction usually refers to a decrease in the number of flower parts within one floral series. Numerous flower parts is considered unspecialized, and fewer flower parts is considered specialized. Ornamentation refers to numerous and diverse specializations in the morphology of flower parts. Therefore, relatively unspecialized flowers would have little or no fusion, numerous separate flower parts in each floral series, and little or no ornamentation. Relatively specialized flowers would exhibit fusion, fewer flower parts in each floral series, and some ornamentation.

Color the diagrams of the flower parts of a relatively unspecialized flower, ranunculus, at the top of the plate. The dots indicate that numerous flower parts may be present in one or more floral series.

A ranunculus flower is relatively unspecialized, usually having a calyx with five separate or distinct sepals. Some unspecialized flowers have numerous sepals. If there is no fusion between the sepals of a flower, the calyx is *aposepalous* (meaning "apart sepals"). Another term used for distinct sepals is "polysepalous." The corolla of ranunculus may contain from five to numerous distinct petals. If there is no fusion between flower petals, the corolla is *apopetalous* (meaning "apart petals"). Another term used for distinct petals is "polypetalous." The androecium of ranunculus contains numerous (ten or more) *distinct* stamens. No specific term is applied to this condition. The ranunculus gynoecium has numerous distinct simple pistils (each consisting of a single carpel). Since the simple pistils, or carpels, are all separate, the gynoecium is *apocarpous* (meaning "apart carpels"). Another term used for distinct simple pistils is "polycarpous."

Color the diagrams of the specialized phlox flower.

A phlox flower is relatively specialized. The calyx usually has five sepals fused into a single unit. If sepals are fused even slightly, the calyx is *synsepalous* (meaning "together sepals"). Another term for fused sepals is "gamosepalous." The corolla of phlox consists of five petals fused into a funnel-shaped floral tube. If flower petals are fused, or connate, the corolla is *sympetalous* (meaning "together petals"). Another term for fused petals is "gamopetalous."

In phlox the lower portion of the filament of each stamen is fused, or adnate, to the flower tube so that each stamen appears attached to the corolla. Stamens that are so fused are called *epipetalous* (meaning "upon petal." The gynoecium of the phlox flowers has a compound pistil formed by the fusion of three carpels. If a flower has a compound pistil, the gynoecium is *syncarpous*. Reduction in the phlox flower is apparent by the small number of flower parts in each flower series.

Color the diagrams labeled "ornamentation" at the bottom of the plate. The dashed lines indicate planes of symmetry.

Specialization of flower symmetry usually involves adaptations to position the pollen carrier for efficient pollination. *Regular* (actinomorphic) flowers, those with radial (wheellike) symmetry, are considered less specialized than *irregular* (zygomorphic) flowers. In *regular* flowers, such as phlox, the size, shape, and relative position of all flower parts within a floral series is the same. The flower parts in one or more floral series differ in size, shape, or relative position in *irregular* flowers, such as sweet pea. The corollas of a phlox flower and a sweet pea flower are illustrated.

Another example of ornamentaion is the specialization of the shape, size, position, color, or other morphological features of flower parts. As an example, the relatively unspecialized flowers of ranunculus have *laminar* (leaflike) *petals;* those of columbine flowers are highly specialized and not laminar. In columbine flowers, the petals form tubular, nectar-producing *spur petals*.

FUSION, REDUCTION AND ORNAMENTATION.

UNSPECIALIZED ★

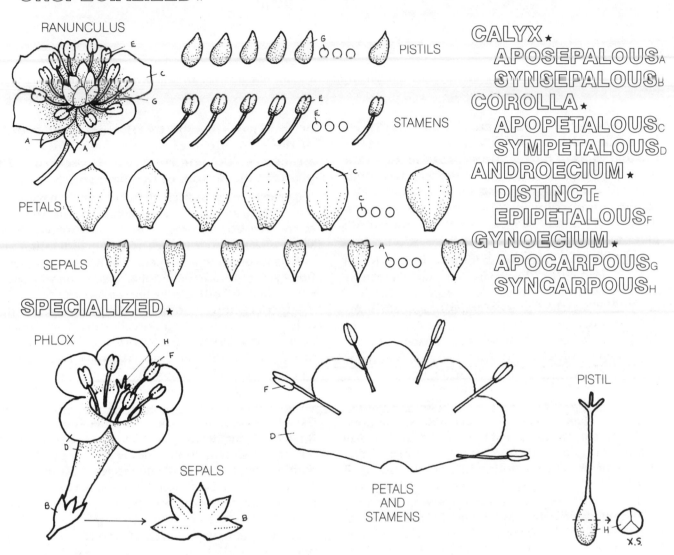

RANUNCULUS

PISTILS

STAMENS

PETALS

SEPALS

CALYX ★
  APOSEPALOUS A
   SYNSEPALOUS B
COROLLA ★
  APOPETALOUS C
   SYMPETALOUS D
ANDROECIUM ★
  DISTINCT E
  EPIPETALOUS F
GYNOECIUM ★
  APOCARPOUS G
  SYNCARPOUS H

SPECIALIZED ★

PHLOX

SEPALS

PETALS
AND
STAMENS

PISTIL

X.S.

ORNAMENTATION ★

FLOWER SYMMETRY ★

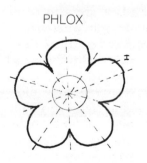

PHLOX

SWEET PEA

REGULAR I

IRREGULAR J

PETAL MODIFICATION ★

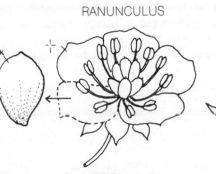

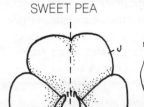

RANUNCULUS

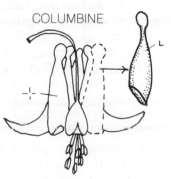

COLUMBINE

LAMINAR
PETAL K

SPUR PETAL L

90

ENHANCING CROSS-POLLINATION

Color the labeled receptacle (A), calyx (B), corolla (C), androecium (D), and gynoecium (E) on the diagrams of perfect flowers at the top of the plate. Be sure to color only the labeled structures in each diagram and leave the unlabeled structures uncolored. Also color the bisexual flowers (F) in the diagram of the monoclinous plant at the left.

Though the basic ground plan for most flowers is four floral series—*calyx* (the sepals), *corolla* (the petals), *androecium* (the stamens), and *gynoecium* (the pistil or pistils)—attached to a common *receptacle,* many variations due to the absence of one or more of the floral series exist as adaptations to enhance effective cross-pollination (pollen transfer between flowers on different plants), efficient pollination, or both.

Perfect, or *bisexual,* flowers always have both fertile floral series, the *androecium* and the *gynoecium,* present. Perfect flowers may have both sterile floral series present, the *calyx* and the *corolla,* or they may lack one or both sterile floral series. "Perfect" refers only to the presence of both fertile floral series. If all four floral series are present, the flower is said to be complete. If one or more floral series, whether sterile or fertile, are absent, the flower is said to be incomplete. Thus, a complete flower is always perfect, since all floral series are present, but a perfect flower is incomplete if it lacks one or both sterile floral series. When an incomplete perfect flower lacks only one sterile floral series, the *corolla* is usually absent, even though the sepals may appear quite petallike in color and texture. The absence of one or both sterile floral series is usually an adaptation for increased pollination efficiency, where wind is the pollen carrier for transferring pollen from one flower to another. Plant species with perfect flowers are called monoclinous since both sexes, male and female, are present in each flower.

Color the labeled structures on the two rows of diagrams illustrating the various arrangements of floral parts in imperfect flowers in the middle of the plate. Be sure to color only the labeled structures. Do not color the flowers on the plants at the bottom of the plate yet.

Imperfect, or unisexual, flowers always have only one of the fertile floral series, either the *androecium* or the *gynoecium,* present. Since at least one floral series is lacking, one of the two fertile series, they are always incomplete. As in perfect flowers, both sterile floral series, the *calyx* and the *corolla,* may be present, or one or both may be absent.

Imperfect flowers in which the *gynoecium* is the only fertile series present are called pistillate flowers because only pistils, one or more, are present. Those in which the *androecium* is the only fertile series present are called staminate flowers since only stamens are present. The absence of one of the fertile series is usually an adaptation to ensure outbreeding. Gamete transfer must take place between different flowers, staminate and pistillate, because pollen cannot be transferred from *androecium* to *gynoecium* within a single flower. Plant species with imperfect flowers are called diclinous because the two sexes are in two separate flowers.

Color the male (G) and female (H) flowers on the two types of diclinous plants labeled "monoecious" and "dioecious," at the bottom of the plate.

Diclinous plant species have various arrangements, usually consistent within a plant species, of staminate *(male)* and pistillate *(female)* flowers. Plant species with unisexual flowers of both types, staminate and pistillate, present on a single plant are said to be monoecious (meaning "one home"), in reference to the occurrence of both staminate and pistillate flowers on the same plant. Plant species with unisexual flowers in which the staminate and pistillate flowers are distributed on separate plants are said to be dioecious (meaning "two homes"), in reference to the occurrence of staminate and pistillate flowers on different plants of the same species.

SEXUAL VARIATIONS.

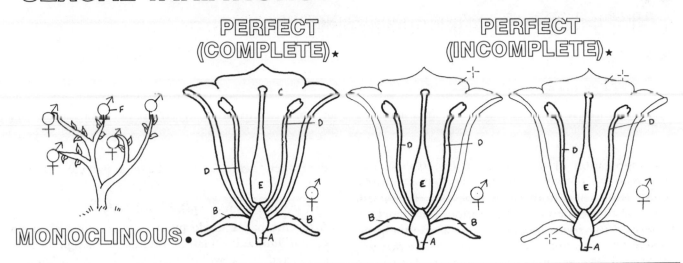

PERFECT (COMPLETE) ★

PERFECT (INCOMPLETE) ★

MONOCLINOUS.

RECEPTACLE A
CALYX B
COROLLA C
ANDROECIUM D
GYNOECIUM E

IMPERFECT (INCOMPLETE)—PISTILLATE ★

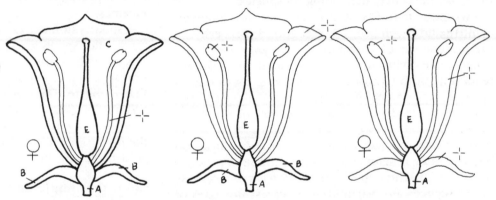

IMPERFECT (INCOMPLETE)—STAMINATE ★

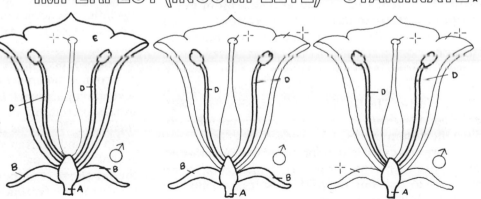

FLOWERS.
BISEXUAL (PERFECT) F

UNISEXUAL (IMPERFECT) ★
MALE G
FEMALE H

DICLINOUS.
MONOECIOUS ★

DIOECIOUS ★

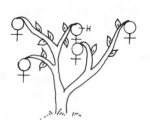

91

THE FLOWERING BRANCH

The aggregation of *flowers,* from one (solitary) to numerous, on a flowering branch is called an inflorescence. Different inflorescence types are recognized, based on the sequence of maturation, the branching pattern of the *stems* (peduncles, etc.), and the presence (pedicelate *flowers*) or absence (sessile *flowers*) of *pedicels. Bracts,* small modified leaves that may be present immediately below *flowers* (whether sessile or pedicelate) or branches, may or may not be present in an inflorescence, depending on species.

Two basic patterns of inflorescence development, cymose and racemose, are recognized, primarily by their sequence of maturation. In cymose inflorescences, the main *stem* is terminated by the first-blooming *flower* so that no further growth occurs from its tip. Lateral branches from the main *stem,* each terminated by a single *flower,* produce the second-blooming *flowers,* and so on. In racemose inflorescences, the main *stem* is terminated by an apical bud that may remain active and continue to produce new growth. The first-blooming *flowers* in racemose inflorescences are usually those nearest the bases of branches.

Color the dichasium cyme and scorpioid cyme at upper left. Note that flowers are represented by circles, and the larger the circle, the older the flower.

A dichasium cyme is a balanced cymose inflorescence. Each branch has a single node from which two lateral branches arise. The *flowers* may or may not have well-developed *pedicels.* Strawberries and cinquefoils produce dichasium cymes.

A scorpioid cyme is an unbalanced cymose inflorescence formed by the consistent formation of lateral branches to one side only. The result is a coiled inflorescence having a scorpion tail or fiddle-neck appearance. The youngest *flowers* are at the center of the coil, and as maturation progresses, the scorpioid cyme uncoils to permit flowering and fruit formation. The *flowers* typically are sessile. Comfrey and forget-me-nots produce scorpioid cymes.

Color the panicle at upper right.

A panicle is a racemose inflorescence that is highly branched. The main *stem* is usually terminated by an *apical bud.* The *flowers* typically have well-developed *pedicels.* Oats have a panicle inflorescence.

Color the raceme, spike, and corymb.

A raceme has a single main *stem* terminated by an *apical bud,* and no lateral branches. A single *flower,* supported by a well-developed *pedicel,* originates from each node along the *stem.* Due to variations in *pedicel* and internode lengths, a raceme may be open or compact. Foxglove and most mustards produce racemes.

A spike is similar to a raceme except that the *flowers* are sessile. Spikes are typically narrow and elongate.

A corymb is similar to a raceme except that the inflorescence is flat topped (convex to concave) rather than elongate. The *flowers* have well-developed *pedicels* and originate singly from nodes along the main *stem,* but due to the unequal length of the *pedicels,* the *flowers* are all held at about the same level. Some goldenrods produce corymbs.

Color the simple umbel and compound umbel.

A simple umbel has a single main *stem* that lacks internode elongation. Thus, the well-developed *pedicels* originate from one common position. In most umbels, the inflorescence is dome shaped. *Bracts* usually form a ringlike structure, called an *involucre,* below the attachment point of the *pedicels.* Onions and milkweeds produce simple umbels.

A compound umbel is an umbellate arrangement of several simple umbels or umbellets. The ring of *bracts* usually present below each simple umbel is called an *involucel.* The *stems* that support the simple umbels are in an umbel arrangement and are called *rays.* The ring of *bracts* usually present below the *rays* is called the *involucre.* Compound umbels are found only in the carrot family, Umbelliferae, which means "umbel bearing."

Color the rest of the plate.

A head, or capitulum, is an inflorescence in which sessile *flowers* are tightly clustered on a concave to convex common *receptacle.* Few to numerous *flowers* may be present within a head. *Bracts* that form an *involucre* are usually present. Members of the sunflower family, Compositae, produce heads that often appear to be a single *flower* even though hundreds of *flowers* may form the flowerlike head.

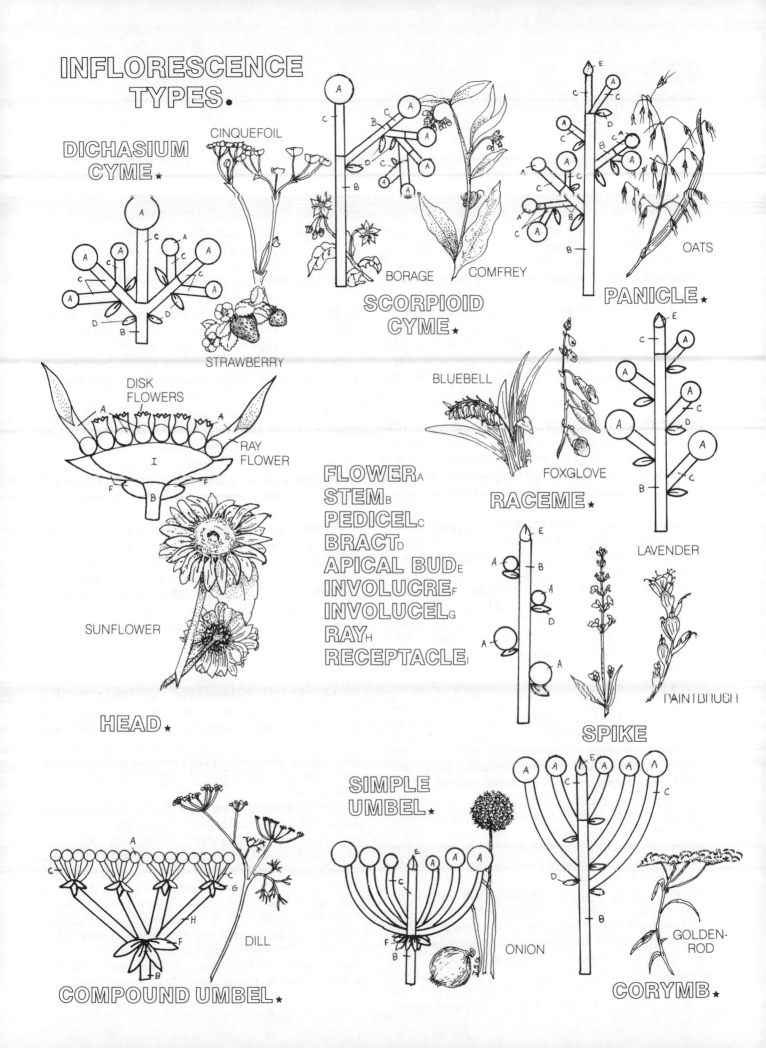

INFLORESCENCE TYPES.

DICHASIUM CYME. ★

CINQUEFOIL

STRAWBERRY

SCORPIOID CYME. ★

BORAGE

COMFREY

PANICLE. ★

OATS

DISK FLOWERS

RAY FLOWER

SUNFLOWER

HEAD. ★

BLUEBELL

FOXGLOVE

RACEME. ★

LAVENDER

FLOWER A
STEM B
PEDICEL C
BRACT D
APICAL BUD E
INVOLUCRE F
INVOLUCEL G
RAY H
RECEPTACLE I

PAINTBRUSH

SPIKE

COMPOUND UMBEL. ★

DILL

SIMPLE UMBEL. ★

ONION

GOLDEN-ROD

CORYMB. ★

The development of adaptations in flowering plants that increase the efficiency of pollen transfer, or pollination, accounts, in part, for their tremendous success. Flowering plants utilize a diverse array of pollen carriers, or *pollen vectors,* including wind, water (rare), and many kinds of animals, especially insects, to ensure successful gamete transfer. Since certain floral features are associated with a particular vector type, predictions as to the *pollen vector* type can be made based on morphological characteristics of a flower.

Color the receptacle (A), calyx (B), corolla (C), stamens (D), pistils (E), and pollen vector (H) on the diagram labeled "wind" at the top of the plate. This is the first of a two-part set. Maintain the same color scheme on the next plate.

Wind-pollinated flowers or "wind flowers" depend upon an inanimate *pollen vector,* air currents, for pollen transfer. Since a large perianth would impede air currents, the *calyx* and *corolla* of these flowers are usually reduced or absent. The *stamens* are often well exserted and have long, narrow filaments that join by a thin connection to the middle of long anthers. This allows them to rock back and forth, releasing pollen with the slightest breeze. Because pollen transfer depends upon the vicissitudes of the *wind,* dry pollen is produced in copious amounts, as it is mere chance that a pollen grain will contact a receptive stigmatic surface of the same species. The *pistils* of wind-pollinated flowers usually have branched styles with long, plumose (featherlike) stigmas that are sticky so that any pollen contacting them will adhere, but only pollen of the same species usually germinates. Style branches are usually well exserted to expose them fully to air currents.

Corn, a good example of a wind-pollinated flowering plant, produces unisexual flowers. Male flowers are in clusters, called tassles, high on the plant. This places them in a favorable position for pollen dispersal by *wind.* The female flowers are lower on the plant, and their long stigmas, called silk, form a cottony mass that effectively catches pollen.

Flowers pollinated by living vectors must have adaptations that attract and position the *pollen vector* for effective pollination. The exquisite structure, background coloration, intricate color patterns, and *fragrances* of flowers have evolved as the result of the co-evolution of flowers and *pollen vectors.* Flower shape, background color, and *fragrance* serve to attract *pollen vectors* at a distance and bring them to the flower; intricate color patterns, internal structure, and placement of the reward (usually food in the form of pollen or *nectar* or both) function to position the *pollen vector* for efficient pollination. In many plants, cross-pollination is enhanced by differences within one flower in the time of pollen release and stigma receptiveness. *Beetles, flies,* butterflies, moths, bees, wasps, and birds are the major animate *pollen vector* types.

Color the floral parts and the indication of fragrance (F) on the diagram labeled "beetle" as well as the beetle pollen vector (H).

Some *beetles* feed heavily on pollen and other floral tissues and in moving from flower to flower in search of a food reward, transfer pollen from anther to stigma to achieve pollination. Beetle-pollinated flowers often have a large, light-colored perianth because *beetles* are poor fliers and need a large surface on which to land. The light-colored flowers are often white or various pastel colors and usually produce a pungent fruity *fragrance,* both of which attract the *beetles* from a distance. Magnolias and water lilies are both good examples of generalized beetle-pollinated flowers.

Color the features of the diagram labeled "fly," including nectar (G) and the fly pollen vector (H).

Many of the *flies,* order Diptera, that function as *pollen vectors* are attracted by sweet *fragrances* and a nectar reward, but *flies* do not consume pollen. A rather wide range of floral variation and relatively unspecialized flowers will attract these nectar-feeding *flies.*

Another reason some *flies* are attracted to the flowers of certain plant species is because the flowers effectively copy rotting meat, usually through a combination of color scheme and odor. These *flies* need to deposit their eggs in rotting materials so the hatched larvae will have a food source. Most plants that have taken advantage of this requirement produce a putrid *fragrance* (odor), which can be overpowering, that attracts *flies* from great distance and a flower color scheme that provides a close-in attraction. *Flies* sometimes actually lay eggs in these flowers, but as there is no real food, the larvae soon die of starvation.

WIND, BEETLES AND FLIES.

FLOWER ★
RECEPTACLE A
PERIANTH ★
CALYX B
COROLLA C
STAMEN D
PISTIL E
FRAGRANCE F
NECTAR G
POLLEN VECTOR H

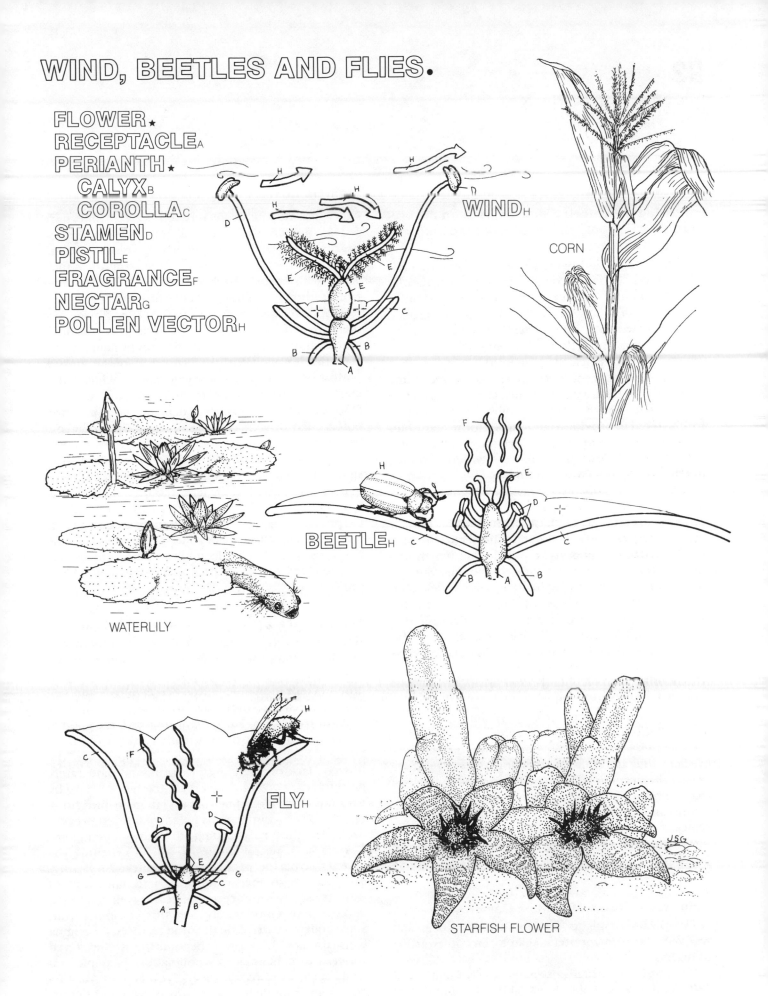

WIND H

CORN

WATERLILY

BEETLE H

FLY H

STARFISH FLOWER

Following the coloring scheme used on the preceding plate, color the features of the butterfly and moth flowers.

Butterfly and moth flowers are often arranged in clusters and usually have a long, narrow, tubular *corolla* or open tubular *corolla* with a narrow pouchlike structure called a nectar spur. *Corolla* background colors are variable, but *butterflies* seem to prefer white, cream, and yellow, pink, and blue pastels. A sweet *fragrance* is usually produced.

Butterflies and most *moths* must land to feed. Many butterfly flowers are upright; the *corolla* flares into a broad surface above the *corolla* tube; and the flowers are clustered to provide a landing platform. Many butterfly flowers have nectar guides (contrasting color patterns on the *corolla*) to orient the *pollen vector* for an effective approach to the flower.

Because these *pollen vectors* have a high energy demand and utilize *nectar* as a major food source, ample *nectar* is secreted at the base of the *corolla* tube or spur. *Butterflies* and *moths* obtain the *nectar* by using a long, narrow proboscis (tongue). The length and narrowness of the *corolla* tube or spur discourages nectar-robbing visits by ineffective, short-tongued insects. The *stamens* are usually included within the flower tube or slightly exserted. The well-exserted stigma enhances cross-pollination since the *pollen vector,* carrying pollen from other flowers, contacts the stigma before the *stamens.* Examples are valerian (*Centranthus*), some phloxes (*Phlox*), and many species of the sunflower family.

A number of *moths* that are active during the evening are able to hover quite well. Flowers pollinated by these *pollen vectors* usually have white to cream colored *corollas* and emit a permeating, sweet *fragrance* during the evening hours. The flower position varies from upright to horizontal. Since the pollen is transferred on the body of the *moth,* the *stamens* are typically well exserted, and the stigma is further exserted. Examples of evening moth flowers are jasmine (*Jasminum*) and many of the evening primroses (*Oenothera*).

Color the features of the bee flower.

More species of plants are pollinated by *bees* and wasps, order Hymenoptera, than by any other group of insects.

Bee flowers are usually open or broadly tubular. *Corolla* background color is variable, but white, yellows, oranges, blues, and purples predominate. Since *bees* apparently perceive pure red as a gray tone, pure red bee flowers are not found. Many flowers that appear red to us have other underlying colors visible to bees. Most bee flowers emit a sweet *fragrance* that, along with *corolla* background color, attracts *bees* from a distance. Though *bees* are good fliers, they cannot hover and must land to feed. Bee flowers are therefore often upright with a flaring *corolla* or, if tubular and horizontal, have a horizontal landing platform that is usually formed by portions of the *corolla*.

Most bee flowers have nectar guides, which often consist of intricate patterns of spots, lines, or both. Since *bees* have a high energy demand and a short proboscis, bee flowers are usually shallow or have a broad entrance if tubular to allow access to the ample *nectar* at the base of the *corolla* or within a short spur. Since *bees* gather pollen from their back and belly by combing with their legs, the *stamens* are adapted, through positioning or movement, to apply ample pollen to the *pollen vector* in these areas. The stigma is typically exserted beyond the *stamens* to facilitate cross-pollination. Examples of bee flowers are orange (*Citrus*), sage (*Saliva*), and some columbines (*Aquilegia*).

Color the features of the hummingbird flower.

Hummingbird flowers are often large and tubular to match the long bill of the *pollen vector.* Reds and yellows are usually a predominant part of their background coloration since birds perceive reds and yellows quite well. Hummingbird flowers are usually without *fragrance.* Since *hummingbirds* are good at hovering, flower position usually ranges from horizontal to pendant (hanging down). The position of the flowers, lack of a landing platform, and moderately deep tubular *corolla* discourages ineffective visits by *bees* and other would-be robbers. In many bird flowers, the ovary is inferior as an adaptation to prevent damage by the bird's probing bill. Many hummingbird flowers have nectar guides to aid in orienting the *pollen vector* for proper flower entry. *Hummingbirds* have high energy demands, and hummingbird flowers usually produce ample *nectar* well concealed at the base of a moderately deep flower tube or spur. Since pollen is carried on the head feathers, the stigma is usually positioned well beyond the well-exserted *stamens* to enhance cross-pollination. Examples include fuchsias (*Fuchsia*), some columbines (*Aquilegia*), some monkeyflowers (*Mimulus*), and hibiscus (*Hibiscus*).

BUTTERFLIES, MOTHS, BEES AND BIRDS.

FLOWER ★
RECEPTACLE A
PERIANTH ★
  CALYX B
  COROLLA C
STAMEN D
PISTIL E
FRAGRANCE F
NECTAR G
POLLEN VECTOR H

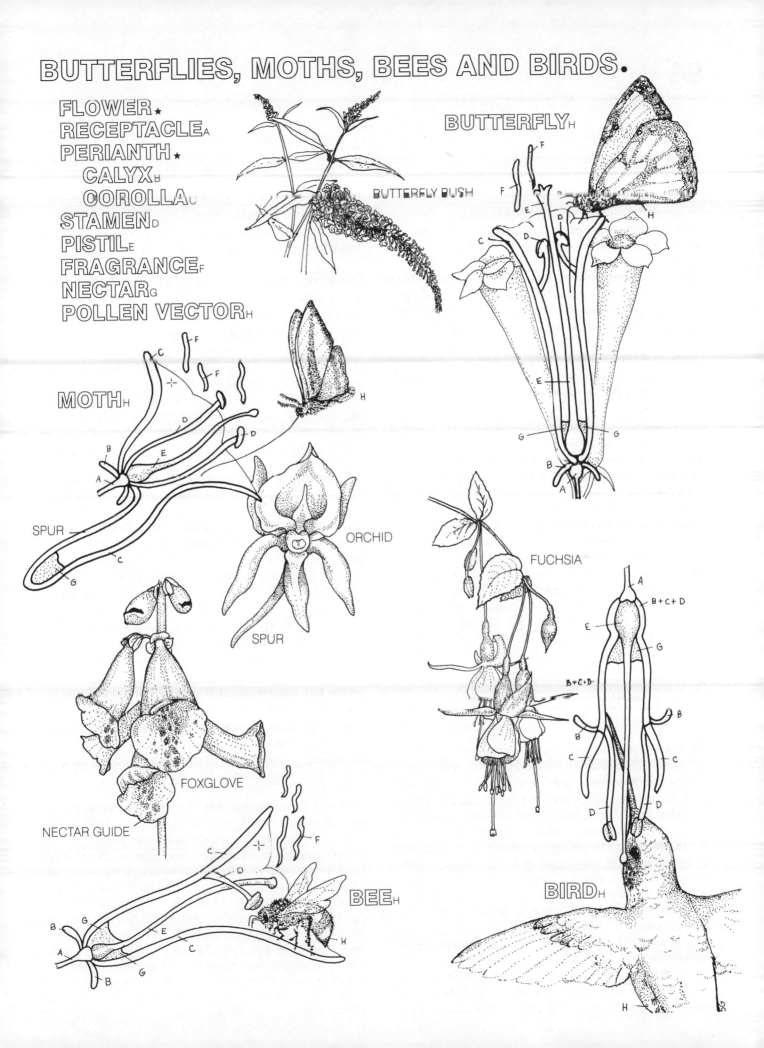

BUTTERFLY BUSH

BUTTERFLY H

MOTH H

SPUR

ORCHID

SPUR

FOXGLOVE

NECTAR GUIDE

BEE H

FUCHSIA

B + C + D

B + C + D

BIRD H

A fruit may be defined as a matured ovary (pistil). In some plant species, the fruit may include other flower parts that are fused to the ovary. True fruits develop from a single pistil, whether simple or compound. A flower with one or more simple pistils produces a corresponding number of fruits. A compound pistil develops into one fruit but may split, at maturity, into a number of pieces that often corresponds to the number of carpels in the pistil. False fruits consist of combinations of two or more simple pistils of a single flower or the pistils of two or more flowers.

Color the achenes (E) and the samaras (F). Color the examples as well.

The *pericarp* or wall of a fruit consists of the mature ovary (pistil) wall and, in some fruits, such as those derived from an inferior ovary, other flower parts as well. The texture of the mature *pericarp,* dry or fleshy, is one characteristic used in fruit classification. Fruits that split open to release their *seeds* are called dehiscent fruits; fruits that remain intact and do not release their *seeds* are called indehiscent fruits.

Dry fruits have a dry *pericarp* and are either single-seeded and indehiscent (this plate) or many-seeded and dehiscent (next plate). The first two dry, single-seeded fruit types described are derived from simple pistils.

Many plant species having flowers with one or more simple pistils, such as many species in the rose (Rosaceae) and the buttercup (Ranunculaceae) families, produce *achenes* (akenes). Each *achene* is derived from one simple pistil and has a single *locule* containing a single *seed* attached only by its *funiculus* to the *pericarp.* Since a simple pistil has a superior ovary, the *pericarp* consists only of pistil tissue. The seed coat is not fused to the *pericarp.* The outer surface of the *pericarp* may be smooth or textured, and it may be covered with hairs, spines, bristles, or hooks that aid in dispersal. The style may fall off, as in cinquefoil *(Potentilla),* or it may be adapted to enhance dispersal. In many buttercups *(Ranunculus),* the style forms a curved beak, or hook; in virgin's bower *(Clematis)* the style elongates during achene maturation to form a long, hairy tail on the *achene.*

The *samara,* as defined here, is similar to the *achene.* The *samara* differs from the *achene* in having a winglike outgrowth from the *pericarp,* which may be on one end, as in the ash *(Fraxinus),* or com-

pletely around it, as in the hop tree *(Ptelea).* Thus, a *samara* may be defined as a winged *achene.*

Color the features of the two fruit types, the mericarp (G) and the schizocarp (H). Color the samples as well.

The remaining dry, single-seeded fruit types are derived from a compound pistil. In the first two types, the carpels that form the compound pistil separate into two individual fruits.

Mericarps are derived from a compound pistil, usually consisting of numerous carpels, that has a superior ovary. In the compound pistil, each carpel forms a one-loculed, one-seeded unit. Though fused during fruit maturation, the carpels typically separate from one another at maturity into individual fruits, called *mericarps,* that are like wedges of a pie. *Mericarps* are common in the geranium (Geraniaceae), maple (Aceraceae), and mallow (Malvaceae) families. *Mericarps* may be untailed, as in hollyhock; tailed, as in stork's bill *(Erodium);* or winged, as in maples *(Acer),* in which the two *mericarps* often remain fused.

A special type of *mericarp* produced by a bicarpellate inferior ovary, the *schizocarp,* is unique to the carrot family, Apiaceae (Umbeliferae). The two one-seeded carpels, each an individual *schizocarp,* separate at maturity.

Color the diagrams of the caryopsis (I), cypsela (J), and nut (K).

Each of the following simple, dry, indehiscent fruits is formed by unilocular compound pistils. The *caryopsis,* or grain, is restricted to the grass (Poaceae) family. A superior, two-carpelled ovary forms a fruit in which the *seed* coat is entirely fused with the *pericarp* and no *locule* space is present. Another family with its own distinctive fruit type is the sunflower (Asteraceae). This fruit type, *cypsela,* is derived from an inferior ovary composed of two fused carpels. The *seed* is connected to the *pericarp* only by the *funiculus,* and some *locule* space remains. The *nut* has a hard, woody *pericarp* derived from two or more carpels. In the nut, the *seed* is free within the *locule* except for the *funiculus* attachment point, and the entire *pericarp* is woody. The nut is usually single seeded. Filberts and acorns are examples of this fruit type.

DRY, SINGLE-SEEDED FRUIT TYPES.

PERICARP A
LOCULE B
SEED C
FUNICULUS D

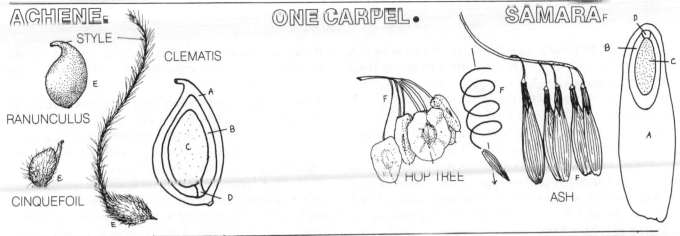

ACHENE E

STYLE

CLEMATIS

RANUNCULUS

E

CINQUEFOIL

E

E

A
B
C
D

ONE CARPEL.

F

F

F

I

HOP TREE

SAMARA F

D
B
C
A

F

ASH

TWO OR MORE CARPELS.

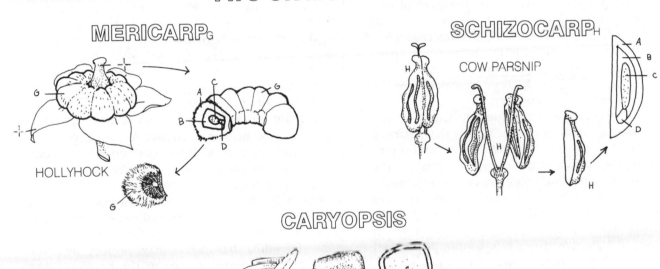

MERICARP G

G

A C
B
D
G

G

HOLLYHOCK

SCHIZOCARP H

COW PARSNIP

H

H

H

A
B
C
D

CARYOPSIS

RICE

I

I

C
A
D

CORN

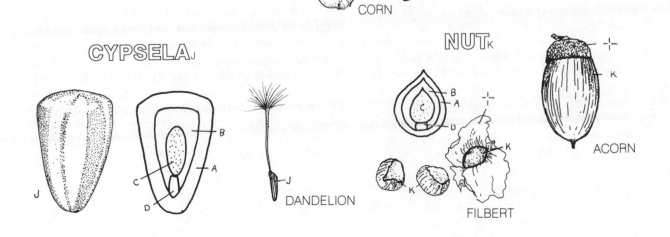

CYPSELA J

B
A
C
D

J

J

DANDELION

NUT K

B
A
C
D

K

K

K

ACORN

FILBERT

Color the pericarp (A), seed (B), funiculus (C), locule (D), and arrows indicating where the fruit splits (E) on the two fruit types, the follicle (H) and legume (I), derived from a simple pistil. Color the examples as well.

Dry, many-seeded fruits, which are derived from a simple pistil or a compound pistil, are dehiscent. Individual *seeds* function as the dispersal unit. It would be an ecological disadvantage if dry, many-seeded fruits were indehiscent and functioned as the dispersal unit because all the *seeds* would be dispersed to the same place. This would result in competition between the seedlings.

As in dry, one-seeded fruits, the *pericarp* of dry, many-seeded fruits is dry at maturity. The fruit has one or more *locules* that may contain numerous *seeds.* Each *seed* is attached to the inner surface of the *pericarp* by a *funiculus.* Though other dry, many-seeded fruit types are recognized, those described are the ones most commonly encountered.

The *follicle* is derived from a simple pistil that contains numerous *seeds.* At maturity, the dry *pericarp* forms a *split* along one side only (the ventral suture formed by the fusion of the carpel margins). In the milkweed, the gynoecium contains two simple pistils. Therefore, each milkweed flower usually produces two separate *follicles.* Other examples of *follicle*-producing flowers are columbine *(Aquilegia),* which produces five *follicles* from five simple pistils, and delphinium *(Delphinium),* which produces three *follicles* from three simple pistils.

The *legume* is a fruit type that is restricted to the pea or legume (Fabaceae or Leguminosae) family. Like the *follicle,* the *legume* is derived from a simple pistil, but it differs from the *follicle* in the formation of two *splits* for dehiscence rather than a single *split.* At maturity, most *legumes* split down both the suture formed by carpel fusion and the suture formed by the carpel midvein. This separates the *pericarp* into two halves, called valves, that fall away or remain attached at their base (near the receptacle). The pea pod, peanut, and bean are all examples of *legumes.*

Color the diagrams of the silicle (J) and silique (J¹), including the replum (F). Color the examples as well.

Two other dry, many-seeded fruit types that are restricted to one plant family are the *silicle* and the *silique.* Both are found in the mustard (Brassicaceae or Cruciferae) family and differ only in length and breadth proportions. The *silicle* is short, being less than twice longer than wide; the *silique* is long, being twice or more longer than wide. These fruit types are derived from a compound pistil, though their exact derivation is unclear. Both are characterized by two *locules* with numerous *seeds,* by a dry pericarp wall that dehisces by two *splits* to create two *pericarp* valves that fall away, and by a membranous partition, called a *replum,* that remains attached to the plant. Stocks *(Mathiola)* produce *siliques;* honesty *(Lunaria)* and candy tuft *(Iberis)* both produce *silicles.*

Color the diagrams of capsule (K) types, including the septa (G). Color the examples as well.

The last dry, dehiscent, many-seeded fruit type covered is the *capsule,* which is derived from a compound pistil. Various types of *capsules,* based on the type of dehiscence, are recognized. The poricidal *capsule,* seen at bottom center, dehisces by several porelike splits that develop near the top of the ovary. The poricidal *capsule* has a single *locule,* though false *septa* (intruding partitions) may project into the *locule* from the inner surface of the *pericarp.* The poppies *(Papaver)* have poricidal *capsules.* Septicidal *capsules* are characterized by *splits* that occur in the *pericarp* along the internal *septa,* the partitions formed by carpel tissue that divide the fruit into two or more *locules.* The separating valves often remain attached at their bases and resemble fused *follicles.* Many members of the lily family have septicidal *capsules.* Loculicidal *capsules* dehisce by *splits* that form in the *pericarp* wall midway between *septa* in the *locule* area. As in septicidal *capsules,* the valves formed by dehiscence of the *pericarp* often remain attached at their bases.

DRY, MANY-SEEDED FRUIT TYPES.

PERICARP A
SEED B
FUNICULUS C
LOCULE D
SPLIT E
REPLUM F
SEPTUM G

FOLLICLE H

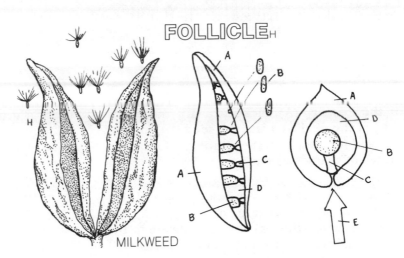

MILKWEED

LEGUME I

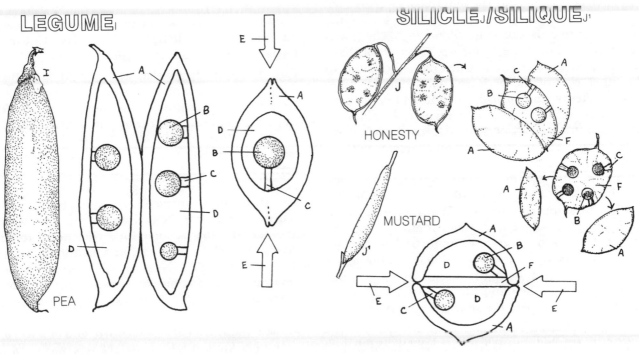

PEA

SILICLE J /SILIQUE J'

HONESTY

MUSTARD

CAPSULE K

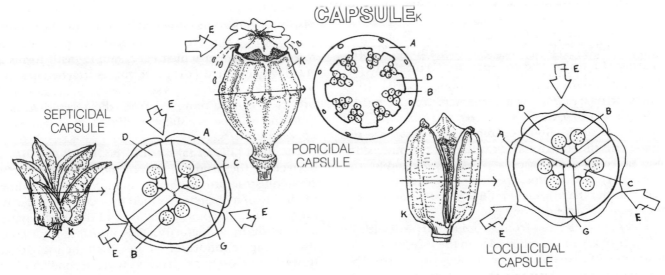

SEPTICIDAL CAPSULE

PORICIDAL CAPSULE

LOCULICIDAL CAPSULE

Color the exocarp (A), mesocarp (B), and endo-carp (C) of the pericarp and the seed (D), funiculus (E), and locule (F) of the drupe (I). Color the examples of drupes as well.

Fleshy fruits, which range from single seeded to many seeded, are characterized by a fleshy fruit wall that often consists of the pericarp, or ovary wall, and other fused attached flower parts. In fleshy fruits, the three layers of pericarp tissue forming the fruit wall can often be recognized. The outermost layer is called the *exocarp*. The middle layer, which is usually the most extensive, is called the *mesocarp*. The innermost layer is called the *endocarp*. Fleshy fruits have one or more *seeds*, each attached by its *funiculus* to the inner wall of the pericarp, within one or more *locules*. Fleshy fruits usually do not dehisce, as their adaptation for dispersal usually involves ingestion by an animal that later disperses the seeds in excrement.

The *drupe* is a simple fleshy fruit characterized by a thin but tough *exocarp*, a fleshy *mesocarp*, and a stoney (hard and woody) *endocarp* that surrounds the single *seed*. The *drupe* is derived from a simple pistil with a superior ovary that produces a single *seed*, though two are sometimes found, within the *locule*. The peach is a good example of a *drupe*. The skin is the *exocarp;* the sweet edible fleshy portion is the *mesocarp;* and the hard "stone" or "pit" is the *endocarp,* which surrounds the enclosed *seed.* When the stone is split open, the *seed* is revealed. Peaches, plums, cherries, nectarines, and almonds are all *drupes.* In the almond, the *seed* is the edible portion and the shell that is discarded is the *endocarp.*

Color the diagrams of the berry (J) and the special berry types, the pepo (K), and the hesperidium (L). Color the receptacle (G) on the pepo.

A *berry* is derived from a compound pistil with a superior ovary. *Berries* are characterized by a thin, skinlike *exocarp,* fleshy *mesocarp* and *endocarp,* and numerous *seeds.* Good examples of a *berry* are the grape and the tomato. The *exocarp* is the outer skinlike layer; the inner fleshy parts are formed by the *mesocarp* and *endocarp.* Two specialized *berry* types, other than the general *berry* just described, are the *hesperidium* and the *pepo.*

The *pepo* is derived from a compound pistil with an inferior ovary that is embedded in the *receptacle.* The leathery, rindlike outer layer of the *pepo* is formed by the *mesocarp* and *endocarp.* Most members of the squash (Cucurbitaceae), such as cantaloupe, squash, and watermelon, have *pepos.* Bananas are another example of a *pepo.*

The *hesperidium* occurs only in the citrus (Rutaceae) family. In the *hesperidium,* the *exocarp* and *mesocarp* form an oil-rich rind, and the *endocarp* forms a white pulp. The compound pistil is divided into several *locules.* Juice-filled, saclike hairs, protruding inward from the *endocarp,* fill each *locule.*

Color the diagrams of the accessory (M), aggregate (N), and multiple (O) fruits. Color the hypanthium (H) on the apple.

False fruits are fruits that are derived from flower parts other than the pistil, from two or more pistils, or from two or more flowers. In one kind of false fruit, the *accessory fruit,* the fleshy part of the fruit is derived from flower parts other than the pistil or accessory flower parts. The fleshy portion of the strawberry *accessory fruit* is the enlarged, fleshy *receptacle* of one flower. The small, gritty, seedlike objects on the strawberry surface are actually individual achenes, which are the true fruits of the strawberry. Another fruit that is sometimes considered to be an *accessory fruit* is the pome, in which the fleshy portion of the fruit is *hypanthium* tissue. A pome develops from a compound pistil with an inferior ovary fused to a *hypanthium.* Apples, pears, and quinces are all examples of pomes. The tough core is formed by the *exocarp, mesocarp,* and *endocarp.*

Another kind of false fruit, the *aggregate fruit,* is derived from a single flower with two or more simple pistils that become attached to one another. Each pistil forms a true fruit. The blackberry produces an *aggregate fruit* in which each simple pistil forms a small *drupe,* called a drupelet, that is attached near its base to adjacent drupelets so the whole aggregation of drupelets can be removed from the *receptacle* as a unit, the blackberry. Many of the "berry" fruits are *aggregate fruits.*

The third kind of false fruit, the *multiple fruit,* is derived by the fusion of two or more flowers into a common structure. In the fig, numerous flowers line the inside of a chamber formed by a common, fleshy *receptacle.* In the pineapple, most of the flower parts become fleshy at maturity. This *multiple fruit* consists of a spike inflorescence in which the individual flowers are closely packed and fused to one another to form a common structure.

FLESHY FRUIT TYPES.

PERICARP★
  EXOCARP A
  MESOCARP B
  ENDOCARP C
SEED D
FUNICULUS E
LOCULE F
RECEPTACLE G
HYPANTHIUM H

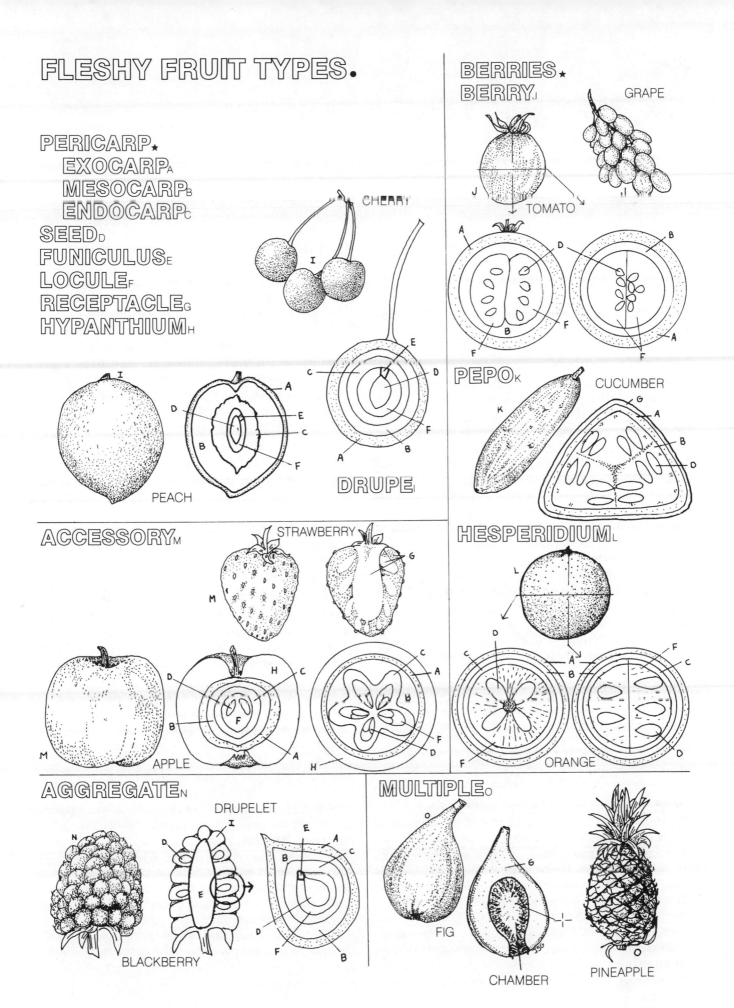

BERRIES ★
BERRY J

GRAPE

TOMATO

PEPO K

CUCUMBER

DRUPE I

PEACH

HESPERIDIUM L

ORANGE

ACCESSORY M

STRAWBERRY

APPLE

AGGREGATE N

DRUPELET

BLACKBERRY

MULTIPLE O

FIG

CHAMBER

PINEAPPLE

97

FLOWERING PLANT DISPERSAL

Since flowering plants are relatively immobile, they depend upon various dispersal units for distribution. In most, the major dispersal unit is the *seed*.

Color the upper left quadrant labeled "expulsion" (D) and the seed (A) in the center. For each dispersal mechanism, color the label and adjacent arrow portions.

Some *fruits* have adaptations that forcibly eject the *seeds* at maturity. Where *expulsion* is the sole means of *seed* dispersal, the distribution of *seeds* is rather localized. Because of this, distributional range increase for a plant species relying solely upon *expulsion* is slow. *Expulsion* creates dense local populations, and competition between individuals for available resources may be severe. One benefit is that most *seeds* are distributed within the habitat type occupied by the *parent plants*; therefore, the chance of landing on a site favorable for germination and successful establishment is high.

Many members of the legume (pea) family produce *fruits* in which pressure within the fruit wall increases as the *fruit* dries and shrinks during maturation. When mature, the two halves of the *fruit* separate and quickly twist with an audible "pop." This action throws *seeds* in all directions about the *parent plant*.

In some *fruits,* such as *Ecballium* and some mistletoes, hydrostatic (water) pressure develops within the fluid-filled ovary chamber. In *Ecballium,* disturbance of the *fruit,* as by touch, causes its immediate release from the pedicel and *expulsion* of the contained *seeds.*

Color the diagrams labeled "wind" (E).

Wind-dispersed *seeds* are distributed by adaptations of the *parent plant, fruit,* or *seed* that utilize air currents as a dispersal vector. Wind functions in both short-distance and long-distance dispersal. Because *wind* dispersal is nonselective, *seeds* are randomly dispersed. Thus, *seeds* land on sites favorable for germination and growth only by chance, and many *seeds* land on unfavorable sites and are wasted. To counterbalance this loss, numerous *seeds* are usually produced.

In tumbleweeds, such as *Salsola, parent plants* function as dispersal units. As a dry, dead *parent plant* tumbles along the ground carried by *wind,* it throws *seeds* in all directions. In maples, *Acer,* and dandelions, *Taraxacum,* the *fruits* function as dispersal units. Maple *fruits* are heavy, but the wings slow their descent and allow for short-distance dispersal. Dandelion *fruits* are lightweight and have a parachutelike structure that increases their bouyancy. Therefore, dandelion *fruits,* which each contain a single *seed,* may be carried for long distances.

Color the diagrams labeled "water" (F).

Fruits and *seeds* that fall into moving *water* may be carried for some distance before becoming lodged downstream. The temporary torrent of a flash flood is a major vector for seed dispersal in some areas (notably deserts). Dispersal units, such as coconut and mangrove *fruits,* may float from the *parent plant* source to distant shores by oceanic currents.

Color the diagrams labeled "animals" (G).

Seed dispersal by *animals* is the most highly adapted and specialized type.

Animals carry *fruits* and *seeds* both externally and internally. Externally transported dispersal units may be carried in the mouth, as acorns are carried by squirrels, or on the fur or feathers, as cockleburs are by a variety of fur-bearing animals. Mouth-carried dispersal units are stashed as reserve food supplies, but many are not recovered. Dispersal units carried on the fur or feathers, whether *seeds* or *fruits,* typically have adaptations that aid in attachment.

Internally carried *seeds* pass through the digestive system. Since this process may take some time, the *seeds* are transported by interim movements of the *animal.* Though many *seeds* are digested, *seeds* adapted for this type of dispersal have hard, impervious, and resistant *seed* coats, and the *seeds* are not harmed during the digestive process. Some *seeds,* such as beavertail cactus (*Opuntia*) and mesquite (*Prosopis*), actually have their germination rate enhanced. The production of edible *fruits,* in general, is an adaptation to ensure *seed* dispersal by *animals.*

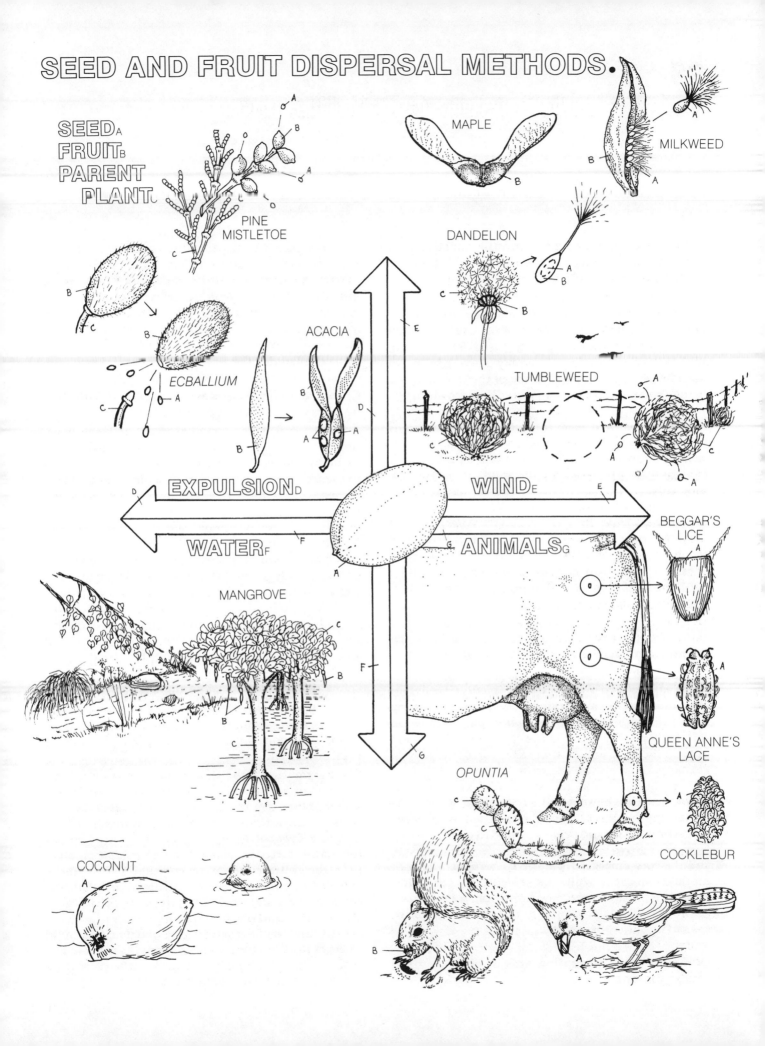

SEED AND FRUIT DISPERSAL METHODS.

SEED A
FRUIT B
PARENT
PLANT C

PINE
MISTLETOE

ECBALLIUM

ACACIA

MAPLE

DANDELION

MILKWEED

TUMBLEWEED

EXPULSION D

WIND E

WATER F

ANIMALS G

BEGGAR'S
LICE

MANGROVE

QUEEN ANNE'S
LACE

OPUNTIA

COCKLEBUR

COCONUT

98

SEEDS AND SEED GERMINATION

Color the diagram of a longitudinal section of a corn kernel. Also color the whole corn kernel.

Corn provides an example of one type of monocot seed germination. It is typical of all grass seeds. Corn, and other grasses such as wheat and rice, form a grain in which the *seed coat* consists of the ovary wall and mature integument, which are fused (not shown). Within the grain is an ample supply of *endosperm* as a nutrient source for maintenance of the dormant, but living, embryo and for early seedling growth upon seed germination. The monocot embryo has a single *cotyledon,* or seed leaf. In grains, the *cotyledon* is closely appressed to the *endosperm.* Enzymes secreted by the *cotyledon* break down the *endosperm* to facilitate absorption and utilization of *endosperm* food reserves by the embryo. Other structures of the embryo are the embryonic *shoot,* which is covered by a tubular, conelike structure called the *coleoptile,* and the embryonic *root,* or radicle, which is covered by a second tubular, conelike structure called the *coleorhiza.* The region between the *root* and *shoot,* called the hypocotyl (not shown), is not extensive and does not elongate appreciably.

Color the first three diagrams illustrating corn germination across the top center of the plate.

Upon germination of the corn grain, the embryonic *shoot,* ensheathed by the growing *coleoptile,* emerges from the *seed coat* and begins growth upward. During early development, as the *shoot* pushes upward through the soil, the *coleoptile* continues to cover and protect the growing *shoot* because *coleoptile* growth keeps pace with *shoot* growth. Embryonic *root* growth begins about the same time as embryonic *shoot* growth. The sheathlike *coleorhiza,* which covers the embryonic *root,* and the enclosed *root* emerge from the pericarp and begin downward growth. However, *root* growth soon surpasses *coleorhiza* growth, and the *root* emerges from the *coleorhiza.* The *root* rapidly elongates while the *coleorhiza* terminates growth. Soon after seed *root* formation, *adventitious roots,* originating from the embryonic *shoot* above the embryonic *root,* emerge.

Color the remaining diagrams of corn seedling establishment.

As development of the seedling continues, the grain shrivels as its *endosperm* is depleted by action of the *cotyledon.* As growth of the *coleoptile* slows, *shoot* growth surpasses *coleoptile* growth and the *shoot* emerges from the tip of the *coleoptile.* In many monocots, seed *root* growth also diminishes, and as *adventitious roots* become well established, the seed *roots* die and deteriorate.

Color the diagram of a whole bean seed and the section at the bottom of the plate.

A bean seed provides a good example of the germination of a dicot seed in which the reserve food is stored in the two *cotyledons.* These are quite evident in peanut *seeds* as the two halves of the edible "nut." Some dicots, such as morning glorys, store nutrient reserves in *endosperm* and have thin, absorptive *cotyledons.* External features of the bean seed are the *seed coat* (the sclerified integument), the *hilum* (the scar of the attachment point of the funiculus), the *micropyle* (the pore in the integument where the pollen tube entered), and the *raphe* (a small ridge on the opposite side of the *hilum* from the *micropyle* where the stalk of the funiculus continued along the ovule). The embryonic sporophyte contained within the *seed coat* consists of an embryonic *shoot* (the epicotyl), two food-storing *cotyledons,* a region of transition from *shoot* to *root* below the *cotyledons,* called the *hypocotyl,* and the embryonic *root* (the radicle).

Color the remaining diagrams of bean seedling establishment.

Upon germination, the embryonic *root* absorbs large amounts of water, swells, and breaks the *seed coat.* The *root* then emerges from the *seed coat* and grows downward. In beans, as the *root* continues elongation, the *hypocotyl* begins to elongate and bend to form an inverted, U-shaped structure that emerges from the soil surface. With continued growth, the *hypocotyl* elongates and straightens to pull the *cotyledons* and *shoot* (epicotyl) free of the disintegrating *seed coat* and out of the soil.

Early growth of the seedling is rapid due to the food reserves stored in the two *cotyledons.* As the *root* and *shoot* systems continue development, the *cotyledons* shrivel as their food supply is depleted.

MONOCOT AND DICOT SEEDLINGS.

SEED COAT A
ENDOSPERM B
COTYLEDON C
COLEOPTILE D
SHOOT E
ROOT F
COLEORHIZA G
ADVENTITIOUS ROOT H

HILUM I
MICROPYLE J
RAPHE K
HYPOCOTYL L

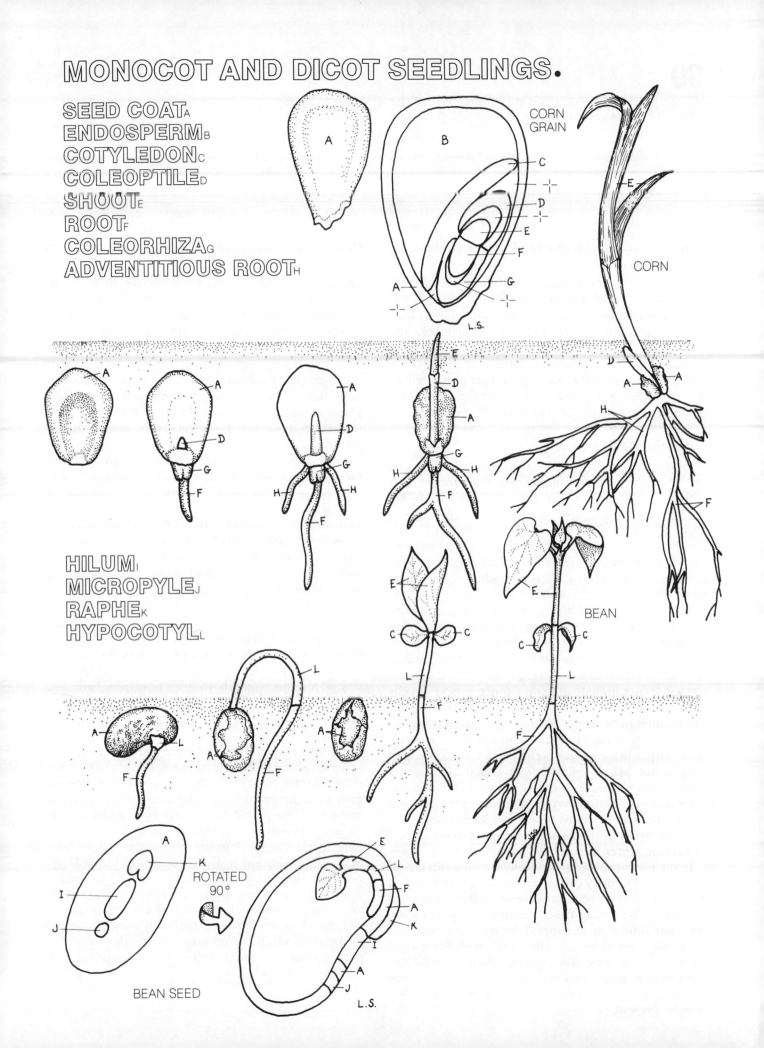

CORN GRAIN

L.S.

CORN

BEAN

ROTATED 90°

BEAN SEED

L.S.

99

ADAPTATION TO THE LAND

Color the marine (A) and land (B) headings at the bottom of the plate.

Many environmental features of the *land* environment differ significantly from the *marine* environment. For example, diurnal (daily) and seasonal fluctuations in temperature, moisture availability, and salinity are slight or nonexistent in the *marine* environment, but these factors usually fluctuate over broad ranges on *land*. Many problems presented by the harshness of the *land* environment must be overcome by adaptations for successful life on *land*.

Color the algae at left.

Most attached algae have an upright growth form and one or more expanded surfaces, called *laminas,* that increase surface area for photosynthesis, *gas exchange,* and *water and mineral uptake*. In most algae, almost all cells are exposed to light and function in photosynthesis. In addition, oxygen and carbon dioxide for both respiration and photosynthesis are found dissolved in the surrounding water. Since algae are completely bathed in water, *gas exchange,* which requires a moist surface, and *water and mineral uptake* occur over the entire algal surface by simple diffusion. Many algae, especially those in the intertidal area, have a *gelatinous layer,* consisting primarily of polysaccharides, covering their surface, but this layer does not impede diffusion. Since almost all cells of an alga manufacture their own food and the plant is surrounded by water, a conduction system is not required for transport of water and nutrients in algae. Some larger brown algae (kelps) that are anchored in deep water below optimal light penetration have a conduction system to transport food downward but no water conduction system. Except for the *gelatinous layer,* which retains water and diminishes water loss during exposure, algae have no special adaptation to prevent water loss.

In the *marine* habitat, most physical support is provided by the density of the surrounding water. Strong and rigid supportive tissues are not adaptive and are not found. To anchor attached forms securely, a *holdfast* that functions in support by adhering to a substrate, without penetrating it, is sufficient. Some algae have a narrow, stalklike portion, called the *stipe,* between the *holdfast* and *lamina* or laminas. To maintain attached forms upright, one or more *gas bladders* may be present.

Color the generalized vascular land plant at right.

Compared to the homogenous water bath of the *marine* environment, the atmosphere of the *land* environment is extremely dry, and *water and mineral uptake* is usually from the substrate. Unlike algae, both photosynthetic and nonphotosynthetic cells are found in land plants, but photosynthesis is largely confined to special cells within *leaves*. *Gas exchange* does not occur over the entire surface since land plants are covered by an *epidermis* and a waxy *cuticle* that form a surface layer relatively impervious to water and atmospheric gases. In land plants, *gas exchange* is confined to internal cells having a moist surface. Pores, through the *cuticle* and *epidermis,* formed by the *stomatal apparatus,* control *gas exchange* and *water loss* by closing during periods of moisture stress and at night in most land plants. Because much of the land plant body is exposed to a relatively dry atmosphere and is covered with the *epidermis* and *cuticle* to diminish *water loss, water and mineral uptake* is typically through the porous *root* (tip) areas located within a moist substrate.

Since many cells are nonphotosynthetic and the photosynthetic cells are some distances from the *roots* that function in *water and mineral uptake,* land plants require a conduction system. *Xylem* carries water and minerals to both photosynthetic and nonphotosynthetic cells, and *phloem* carries food materials from the photosynthetic cells to the nonphotosynthetic cells.

The most successful land plants have taken advantage of the area above the substrate surface to increase their surface area. Physical support for upright growth is not provided by the atmosphere. To overcome this, many land plants have evolved an upright *stem,* containing strong support tissues, to provide a multilevel framework for the photosynthetic surfaces *(leaves)*. In addition to their *water and mineral uptake* function, *roots* also solidly anchor the upright *stem* in the substrate. In woody land plants, secondary growth, through the activity of a *vascular cambium,* provides a means of continually increasing *stem* diameter through the addition of woody tissue that provides the additional support needed to permit tall, upright *stems*.

LIVING ON THE LAND.

LAMINAc
GAS EXCHANGEd
WATER AND
   MINERAL UPTAKEe
GELATINOUS LAYERf
HOLDFASTg
STIPEh
GAS BLADDERi
LEAFj
EPIDERMISk

CUTICLEl
STOMATAL APPARATUSm
WATER LOSSn
ROOTSo
XYLEMp
PHLOEMq
STEMr
VASCULAR CAMBIUMs

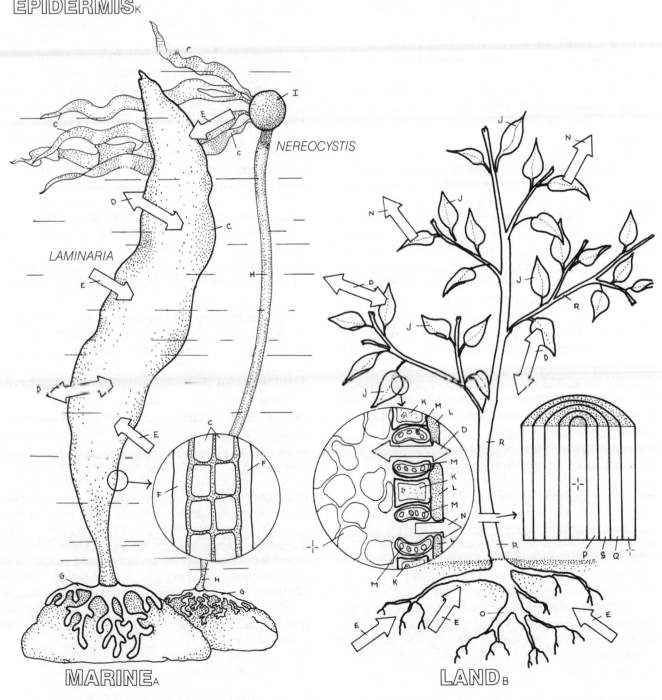

NEREOCYSTIS

LAMINARIA

MARINEa   LANDb

The bryophytes, ferns, conifers, and flowering plants exhibit a sequence of increasing specialization for life on land.

Color the features of the first row of diagrams, depicting gamete transfer.

Gamete transfer, the union of *sperm* and *egg,* in algae requires *water.* One or both gamete types are usually motile.

In bryophytes and ferns, *eggs* are nonmotile and retained by the gametophyte, but both groups produce motile *sperm* that are freely released when *water* is available.

Conifers and flowering plants also retain nonmotile *eggs,* but the *sperm* are nonmotile and do not require external *water* for gamete transfer. In both these groups, *pollen grains,* which contain the male gametophyte, rather than gametes are transferred. The highly resistant *pollen grains* have a long life expectancy and resist desiccation. In conifers, *wind* is the transfer vector. In many flowering plants, *pollen grains* are more directly and efficiently transferred to receptive female structures by *animal* vectors.

Color the features of the second row of diagrams, depicting adaptations for egg protection.

Most algae release both *sperm* and *eggs,* but some retain the *egg* within the wall of the cell, or *oogonium,* that produced it. Thus, some protection may be provided by a cell wall in some algae, but no sterile jacket of protective cells is found.

Land plants produce a sterile jacket of cells called the *archegonium,* which surrounds and protects the *egg.* Fertilization, as well as early embryo development (not shown), takes place within the *archegonium.* For fertilization to occur in bryophytes and ferns, the *archegonium* must be covered with a film of *water* while it is receptive to *sperm.*

In conifers and flowering plants, a four-layered envelope consisting of the *archegonium* formed by the female *gametophyte,* the female *gametophyte,* and the *nucellus* and *integument* formed by the *sporophyte* surround and protect the *egg* from dessication and mechanical injury. In flowering plants, an additional protective layer is provided by the *ovary wall.*

Color the features of the third row of diagrams depicting major dispersal units.

In most algae, *gametes,* whether motile or nonmotile, may function as dispersal units because they are freely released into open *water.* Most algae also produce *zoospores,* though some produce *aplanospores.*

Bryophytes and ferns require dispersal units that are effectively dispersed in the land environment. Dry, lightweight, wind-dispersed *aplanospores* function as the major dispersal units for both these groups.

Conifers and flowering plants are both seed plants, and *seeds* rather than spores are the major dispersal units. Some conifers depend primarily upon *wind* for limited dispersal. Others, with heavy *seeds* such as pinyon pines, depend on *animals* that gather the *seeds* as a food source for dispersal.

Some flowering plant *seeds* are adapted for very localized distribution; others are adapted for very efficient long-distance dispersal. Some, such as self-planting types, have virtually no dispersal, but many depend on various types of dispersal agents including *wind, water,* and many different kinds of *animals.*

Color the features of the fourth row of diagrams depicting the dominant generation and the arrows at the bottom that indicate land plant groups in which the gametophyte is typically dominant and those in which the sporophyte is dominant.

In the algae, some species exhibit *gametophyte* dominance; some exhibit *sporophyte* dominance; and some exhibit a pronounced presence of both generations. In the bryophytes, the *gametophyte* is the dominant generation and the *sporophyte* is completely dependent upon it. In the ferns, the *gametophyte* is independent, but it is small and short-lived in most. Though the *sporophyte* begins its life dependent upon the *gametophyte,* it soon establishes itself as an independent, relatively large, long-lived plant. In the conifers and flowering plants, the female *gametophyte* develops entirely within *sporophyte* tissue as a totally dependent plant. In seed plants, the embryonic *sporophyte* becomes independent following *seed* germination.

REPRODUCTION AND DISPERSAL.

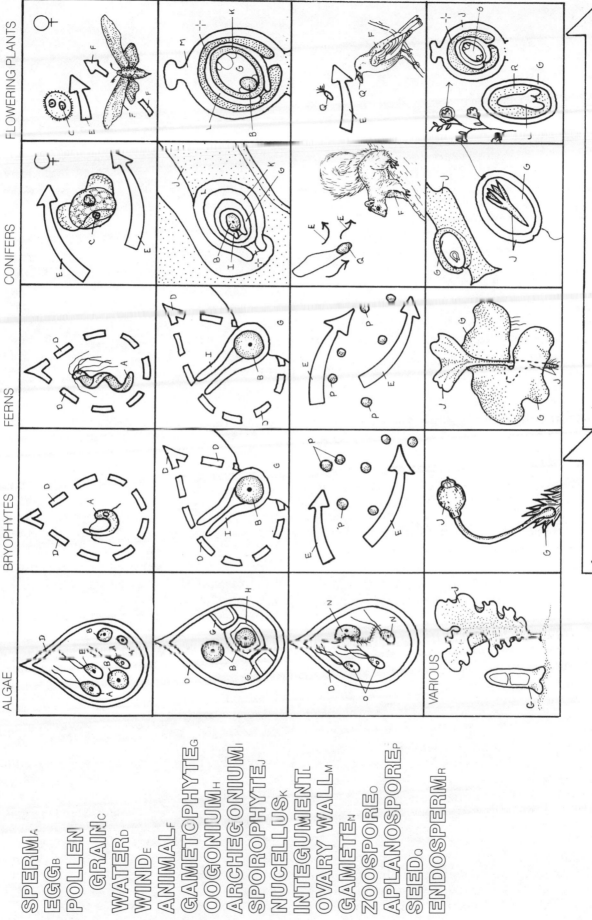

ALGAE   BRYOPHYTES   FERNS   CONIFERS   FLOWERING PLANTS

SPERM A
EGG B
POLLEN GRAIN C
WATER D
WIND E
ANIMAL F
GAMETOPHYTE G
OOGONIUM H
ARCHEGONIUM I
SPOROPHYTE J
NUCELLUS K
INTEGUMENT L
OVARY WALL M
GAMETE N
ZOOSPORE O
APLANOSPORE P
SEED Q
ENDOSPERM R

INDEX